G. Schwedt

Chromatographic Methods in Inorganic Analysis

Chromatographic Methods

Editors

W. Bertsch
W. G. Jennings
R. E. Kaiser

Chromatographic Methods in Inorganic Analysis

Separation Methods and Systems
Detection Methods
Applications in Inorganic Trace Analysis

by G. Schwedt

With 35 Figures and 62 Tables

Dr. Alfred Hüthig Verlag
Heidelberg Basel New York

Prof. Dr. G. Schwedt
Anorganisch-chemisches Institut der Universität Göttingen
Tammannstraße 4
D-3400 Göttingen

ISBN 3-7785-0690-0

Typesetting and printing:
Druckerei Krebs-Gehlen, Hemsbach/Bergstr.
Printed in Germany

Preface

Chromatographic methods of analysis are widely used on a routine basis in the analysis of organic substances in many areas of science, including medicine, and their importance is still increasing*.

In contrast to spectroscopic and electrochemical methods, chromatographic methods are far less commonly used in inorganic analysis. Even though this method of separation may be less suitable for inorganic substances than for organic compounds as a result of their properties, and though the problems of separation encountered in inorganic chemistry are less multifarious, a series of practicable applications utilizing selective and sensitive detector systems has meanwhile been described, particularly in the field of trace analysis.

As a consequence of his traditional training in inorganic analysis, the analyst may frequently have little experience of this technique, and even less knowledge of the potential of chromatographic analysis as applied to his own specific problems.

This book has therefore been written with the explicit aim of providing a comprehensive survey, although the sheer volume of material precludes completeness, of the variety of chromatographic separation systems for a wide range of compounds of the elements (Chapter I), of specialized detection methods used in conjunction with special chro-

---

* See "Chromatographic Methods", a new book series edited by *W. Bertsch, W. G. Jennings*, and *R. E. Kaiser* and published by Dr. Alfred Hüthig Verlag, Heidelberg-New York, which keeps the reader abreast of advances in chromatography.
Hitherto existing titles: *W. Bertsch, S. Hara, R. E. Kaiser, A. Zlatkis*, Instrumental HPTLC (1980); *R. E. Kaiser* and *E. Oelrich,* Optimization in HPLC (1981); *W. Dünges*, Prechromatographic Micromethods (1981); *A. S, Said*. Theory and Mathematres of Chromatography.

matographic techniques (Chapter II), and, above all, of practical applications of chromatographic analysis (Chapter III).

In Chapter III, particular emphasis has been placed on a brief description of sample preparation, on quantitative data, and on a comparison of chromatographic procedures with those used in conventional inorganic analysis. A comprehensive index, arranged according to the periodic classification and according to methods and techniques of chromatography, should permit successful searching for analytical procedures which have already been described and for approaches to the solution of the analyst's own specific problems.

The book assumes a general knowledge of the principles of chromatography, methods and techniques, which can be found in standard textbooks.

Georg Schwedt

Contents

Introduction

Chromatographic methods and organic analysis have become almost synonymous. An analytical procedure for determination of several organic compounds is nowadays hardly conceivable without application of a chromatographic method. However, chromatography provides only a means of separation, but not of determination. Hence a complete analytical procedure requires combination of a chromatographic method with a method of determination (detection).

Analysis of inorganic compounds, generally amounting in practice to elemental analysis, is essentially dominated by the methods of spectroscopy (e.g. photometry or atomic absorption spectroscopy) and electrochemical analysis (e.g. voltammetry/polarography). Since there are far fewer inorganic compounds than organic compounds, the number of separation methods required is also much lower.

Nevertheless, chromatographic methods are also of interest for inorganic analysis and have meanwhile acquired a firm place for a number of specific problems, particularly in trace analysis. Routine analytical laboratories are nowadays generally equipped with gas and liquid chromatographs. Methodological experience gained by an analyst in daily application of these chromatographic techniques to the analysis of organic substances can be profitably applied to chromatographic elemental analysis. Since the choice of an analytical method for a given problem is determined primarily by the experimental know-how that has accrued in a laboratory, chromatographic methods can be successfully applied in particular to those areas in which not only inorganic analysis is carried out.

In a number of cases, i.e. for particular analytical problems (analytical samples or specific elements), the availability of selective and highly sensitive methods of detection (in direct combination with the chromatographic separation process) can sometimes render chromatographic procedures superior to instrumental methods of elemental analysis, such as AAS or voltammetry, with regard both to sample preparation and time requirements and to detection limits and selectivity.

The tasks which chromatographic methods can nowadays fulfil in inorganic analysis can be classified into the following fields:

- Several elements are to be jointly identified and determined in a single procedure (multi-element analysis).
- Interfering substances or, in the case of trace analysis, the entire matrix must be removed prior to application of an element-specific method of determination.
- A small amount of trace element must be enriched from a complex matrix prior to quantitative analysis since the sensitivity of the available method of determination would otherwise be insufficient.
- Very sensitive but hardly selective detectors, such as the flame ionization detector (FID) for gas chromatography or the UV photometer for liquid chromatography, can also be applied to elemental analysis once the elements have been transformed into organic derivatives and the sample mixture is prepared for interference-free detection by the chromatographic separation process.
- Entirely new, special tasks have been opening up in recent years for chromatography in inorganic analysis, now that questions of the chemical fate of (principally toxic) elements (the elemental species) in the environment are being increasingly posed. For example, the analysis of mercury in nature to distinguish between inorganic and organometallic species is now unthinkable without chromatographic methods. Questions regarding the various valence states of an element can frequently also be answered by use of chromatographic separation methods.

Considerable advances have been made in all these areas in the last ten to twenty years, although they cannot always be generalized and often apply only to highly specific problems or even just to individual elements.

Table 1

Monographs on chromatographic analysis.

General textbooks and handbooks

E. Heftman (ed.), "Chromatography. A laboratory handbook of chromatographic and electrophoretic methods", 3rd edit., van Nostrand-Reinhold, New York (1975).
O. Mikes (ed.), "Laboratory handbook of chromatographic and allied methods", Horwood, Chichester (1979).
G. Zweig and *J. Sherma*, "Handbook of chromatography", CRC Press, Cleveland (1972).
G. Schwedt, "Chromatographische Trennmethoden. Theoretische Grundlagen, Techniken und analytische Anwendungen", Thieme, Stuttgart (1979).

(see further references to specialist monographs therein)

Thin-layer chromatography

F. Geiss, "Die Parameter der Dünnschicht-Chromatographie", Vieweg, Braunschweig (1972).
E. Stahl (ed.), "Thin-layer chromatography (A laboratory handbook)", Springer, Berlin-Heidelberg-New York (1969), 2nd edn.
J. Gasparic and *J. Churacek*, "Laboratory handbook of paper and thin-layer chromatography", Horwood, Chichester (1978).
J. C. Touchstone and *M. F. Dobbins*, "Practice of thin-layer chromatography", Wiley, New York (1978).

Column chromatography

H. Engelhardt, "High-pressure liquid chromatography", Springer, Berlin-Heidelberg-New York (1978), 2nd edn.
L. R. Snyder and *J. J. Kirkland*, "Introduction to modern liquid chromatography", Wiley, New York (1980), 2nd edn.

Gas chromatography

R. L. Grob, "Modern practice of gas chromatography", Wiley, Chichester (1977).
R. Kaiser, "Chromatographie in der Gasphase", Vol. I – IV, Bibl. Institut, Mannheim (1969/1973/1975), 2nd and 3rd edn., respectively.
G. Schomburg, "Gaschromatographie", Verlag Chemie, Weinheim (1977).

The greatest expectations were perhaps associated with the possibilities of multielement analysis by chromatographic methods; however, precisely these hopes have been realized least of all. The problems and applications listed demonstrate that chromatographic methods only develop their full potential in combination with other methods and, by this token, are of particular interest for inorganic analysis.

Apart from studies aiming at detection and quantitative determination of various compounds of an element (elemental species), chromatographic procedures frequently necessitate performance of chemical reactions leading to derivatives having more favorable chromatographic properties (chelatization of metals, transformation of anions or their acids into organic derivatives, chemical reactions for detection of ions on thin-layer plates, etc.). This is why the chemistry of the elements commands considerable interest in the development of chromatographic methods.

The following Chapters and their Sections should indicate the state of the art in chromatographic analysis applied to analytical problems of inorganic chemistry with numerous illustrations taken especially from the more recent literature.

Inorganic chromatographic analysis generally shows no significant differences in the execution of the necessary separations from the situation in organic chromatographic analysis. The existing monographs are therefore suitable as introductions to the methods of chromatography and to the individual experimental techniques. A selection of comparatively recent English and German books, some of which also consider inorganic applications, is presented in Table 1.

I Separation Methods and Systems

1 Liquid Chromatography

Compared to gas chromatography (GC) which requires substances volatilizing without decomposition, numerous inorganic substances can be separated in the form of ions or of nonvolatile polar and apolar compounds on use of a liquid as mobile phase.

Depending upon the kind of substance, they can be assigned, primarily with a view to the best possible separations, to the various techniques of liquid chromatography:

- Adsorption chromatography: for metal chelates.
- Partition chromatography: primarily for ions – less suitable for metal chelates.
- Reversed phase chromatography: primarily for ions and ion pairs, recently also for metal chelates.
- Ion-exchange chromatography for simple ions and ionic complexes of elements and for inorganic salts.
- Gel chromatography: for organic complexes of elements and for inorganic salts.

The first successful application of chromatographic separation to inorganic compounds was accomplished with the technique of paper chromatography in the form of partition chromatography of ions. This chromatographic technique has meanwhile been largely superseded by thin-layer chromatography (TLC), which may also utilize cellulose as layer material.

An almost complete survey of these methods with a total of over 700 references up to 1970 (including TLC and column chromatography) is given in the book by *J. Michal*, "Inorganic chromatographic analysis" (1973) [287].

The systems of paper chromatography can be divided into three main groups:

1) Stationary phase: cellulose (+ water)
   Mobile phase: mixtures of organic solvents (especially alcohols and ketones, partly miscible with water such as butanols, etc.) and aqueous solutions of mineral acids, organic acids, bases, organic salts. More than 80 such systems are given for the separation of a total of 45 cations in the Appendix of *Michal*'s book [287].
2) Stationary phase: cellulose impregnated with organic complexing agents (such as 8-hydroxyquinoline, 4-hydroxybenzothiazole, dithizone, etc.)
   Mobile phase: organic solvent mixtures or aqueous systems.
3) Stationary phase: cellulose, impregnated with inorganic or organic ion exchangers such as high-molecular amines (e.g. quaternary ammonium salts) or, e.g., zirconium phosphate
   Mobile phase: mineral acids (or their aqueous solutions).

This is primarily a method of ion-exchange chromatography, with the cellulose serving mainly as support material.

Publications concerned with paper chromatography in inorganic analysis date mainly from the years between 1950 and the mid-1960s, and found their sequel in thin-layer chromatography on cellulose materials. Only the latter approach is considered in the present chapter of this book; some quantitative applications of paper chromatography will be found in Chapters II and III.

Up until the 1960's, liquid chromatography was limited almost entirely (with the exception of ion exchange chromatography) to paper and thin-layer chromatography. The reason is to be seen in the numerous possibilities of performing detection reactions, particularly for ions, after separation on an enormous variety of layer materials, and of eluting the substances of interest from the layer for determination by a wide range of quantitative methods (see Chapter II).

Only in recent years has column chromatography advanced technologically to modern high performance liquid chromatography (HPLC) which is playing a steadily increasing role in inorganic analysis too. This is because HPLC methods permit greater separation performance, shorter analysis times, and above all continuous detection of the column effluent with a variety of highly selective and sensitive detectors (see Chapter II).

Reviews of thin-layer chromatography in inorganic analysis have been written by:

Lesigang-Buchtela, 1966 [252]; *Senf*, 1966 [347]; *Lederer*, 1967 [242]; *Nickless*, 1968 [303]; *Brinkman*, 1973 [50].

The tabular surveys is the following Sections on liquid chromatography are subdivided into thin-layer and column chromatography, depending upon the technique adopted.

1.1 Thin-layer Chromatography (TLC)

1.1.1 Adsorption (TLC)

Chelating agents which have already proved of value for extractive-photometric determinations and for enrichment of traces of elements are also used for thin-layer chromatography of elements.

Intensive studies have been performed for dithiocarbamates and dithizonates (see Table 2). The numerous publications demonstrate not only the versatile methods of separation and their scope (see also Chapter III/2.2), but also the difficulties resulting from the light-sensitivity of the chelates, the formation of several products with one element, incomplete separations, and similar problems.

The analytical significance of such fundamental studies can only be evaluated from the number of practical applications published in the ensuing years. Such applications have been published for specific problems, particularly of a toxicological nature (for detection and quantitative analysis of toxic elements, i.e. heavy metals), so that the chromatography of these metal chelates can be regarded as important, not least because of the first successful separations with HPLC.

Nevertheless, in spite of numerous variants of the mobile phase system, complete separation of all elements reacting with dithizone or substituted dithiocarbamates has not been accomplished.

Table 2

Adsorption chromatography of metal chelates on thin-layer plates.
(In the case of less readily accessible original publications, the number of the table in *Brinkman's* review [50] is given, marked with an asterisk.)

Group of substances	Elements investigated	Reference
Dithizonates	Cu, Hg	*1.03/1.04
	Hg, Pb, Cd	*1.05/1.06
	Cd, Pb, Bi, Cu, Zn, Hg	*1.10
	Zn, Cd, Cu, Pb, Hg	*1.12
	Ag, Au, Pd, Pt	[173]
	Zn, Cu, Ni, Co, Pb, Bi, Cd	[143]
	Hg, Bi, Cu, Zn, Pb, Cd, Ni, Co	*9.18
	Sn, Ag, Tl, Cd, Pb	[336]
	Bi, Cd, Co, Cu, Hg, Mn, Ni, Pb, Zn	[403]
	Pd, Cu, Hg, Cd, Pb, Bi, Co, Ni, Zn	*9.37
	Cu, Pd, Hg, Co, Ni, Pb, Bi, Zn	*9.38
	Cd, Ni, Pb, Co, Cu, Mn, Zn, Hg	[410]
	Cd, Pb, Cu, Ni, Co, Zn	[241]
	Hg, Ag, Bi, Cd, Cu, Pb, Ni, Co, Zn, Mn	*11.61/11.62
	Ag, Pd, Cu, Hg, Zn, Ni, Co, Cd, Tl, Mn, Sn, Fe	*11.87
Dithiocarbamates (see Fig. 1)		
Diethyl-	Pb, Bi, Cd, Cu, Hg	[348]
Diethyl-	Ag, Bi, Cd, Co, Cu, Hg, Ni, Pb, Zn	[65]
Diethyl-	Ag, Hg, Cu, Pb, Bi, Tl, Co, Ni, Cr	[328]
Diethyl-	Ga, In, Tl, Zn, Cd	[329]
Diethyl-	Bi, Cu, Ni, Co, Fe, Mo, U, V, Cr	[216]
Diethyl-	Au, Se, Te, Pt, Pd	[330]
Diethyl-	Ag, As, Bi, Cd, Co, Cr, Cu, Fe, Hg, In, Mn, Ni, Pb, Pd, Sb, Sn, Te, Tl, V, Zn	[13] and *11.76
Diethyl-	Hg, Ag, Bi, Cd, Pb, Cu, Ni, Co, Zn, Mn, Cr, Fe, Sb, Sn	*11.61/11.62
Tetramethylene-	Sn, As, Sb	[350]
Dibenzyl- and diethoxyethyl-	Zn, Cd, Cu, Ag, Pb, Bi, Fe, Co, Ni, Hg	[221]
Various substituents (polychlorinated)	Fe, Co, Ni, Cu, Zn, Cd, Pb, As, Sb, Bi, Se	[391]
Dithizonates and diethyldithiocarbamates	(Theory of adsorption processes)	[126]
Xanthates	Cu, Ni, Co, Mo, Bi, Pb, Zn	[331]
–, fluorinated	Mn, Fe, Co, Ni, Cu, Zn, Cd, Hg, Pb, Bi	[390]
Pyridyl(azo-)naphthol (PAN) complexes	Co, Cu, Ni	[417]
	Cu, Ni, Co, Fe	[125]
	Co, Cu, Ni, Fe	*5.08/5.09
	U, Co, Cu, Ni, Fe, Mn	[100]

Table 2 (Continued)

Group of substances	Elements investigated	Reference
1-Nitroso-2-naphthol complexes	Ni, Fe, Cu, Co Mn, Fe, Co, Ni	[317] [349]
Pyridin-2-aldehyde 2-quinolylhydrazone complexes	Fe, Ni, Cu, Co	[111]
1,2-Diacetylbis(thiobenzhydrazone) complexes	Hg, Pb, Zn, Cd	[161]
Glyoxal bis(4-biphenyl)thiosemicarbazone complexes	Zn, Cd, Hg, Pb, Cu	[304]
Salicylaldoxime complexes	Fe, Co, Ni, Cu, Mn, Bi	[351]
Acetylacetonates	Co, Mn, Ni, Cu, Cr, Fe	[157]
Thenoyltrifluoroacetylacetonates (see Fig. 2)	Cu, Ni, Co	[192]
Monothio-β-diketonates	Cu, Co, Ni, Pd, Rh	[294]
Thiocarbonato complexes	Cu, Ni, Co	[192]
Dithiolates	Pb, Sn, Sb, Bi	[118]
8-Hydroxyquinolates	Bi, Cd, Pb, Cu, Ni, Co, Zn, Mn, Cr Fe, Al, Ca, Mg	*11.61/ 11.62
Survey for following ligands: 8-Hydroxyquinoline, 2-methyl-8-hydroxyquinoline, 7-alkyl-8-hydroxyquinoline, 7-(2,3-dibromopropyl)-8-hydroxyquinoline, cupferron, thionalide, diacetyldioxime, 5-p-dimethylaminobenzylidenerhodamine		*11.35
8-Mercaptoquinoline complexes	Fe, Co, Ni, Ru, Rh, Pd, Os, Ir, Pt Cu, Zn, Hg, Sn, Pb Fe, Co, Ni	 [222] [223]
1-Hydroxy-2-pyridinethione complexes	Ni, Pd, Pt, Rh, Ir	[223]
Alkylaryltriazenes	Ag, Hg, Cu	*1.18
Tetraphenylporphyrin complexes	Fe, Co, Mn, Ni, Zn, Co, Rh, Pb, Cd, Hg	[176]
Tropolone complexes	Ru, Rh, Pd, Ti, V, Fe, Ni, Co, U, Pb	[191]

With dithiocarbamates, variations of the substituents (tetramethylene, dibutyl, diethoxyethyl, or fluorinated alkyl groups instead of diethyl) have in some cases improved the scope of separations and the chromatographic properties in general (reduction of tailing, etc.).

Other chelating agents of interest for TLC separations of elements, such as 8-hydroxyquinoline, acetylacetone, etc. have been collected in Table 2. Silica gel serves as stationary phase in most cases, while alumina has also been used on occasions. The review by *Brinkman*, 1973 [50], presents a exhaustive compilation of chromatographic conditions and Rf values (whenever published) from the literature published up to 1972.

Figures 1 and 2 give examples of successful separations of individual elements with the aid of adsorption TLC of metal chelates. Detailed information concerning chromatographic conditions, the formation of the chelates, and evaluation of thin-layer chromatograms should illustrate the methodology and scope of these techniques.

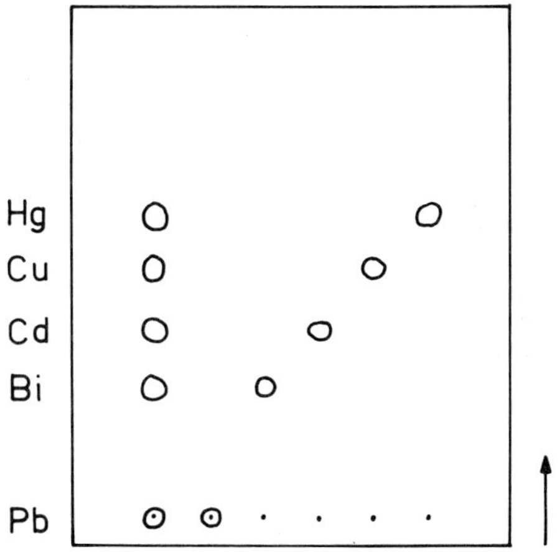

Figure 1

TLC separation of metal diethyldithiocarbamates (DDTC complexes), according to *Senf* [348].
Stationary phase: silica gel D (VEB Chemiewerk Greiz-Dölau, GDR).
Mobile phase: n-hexane/chloroform/diethylamine (20:2:1, v/v).
Length of run: 9 cm. Duration of run: 20 min.
Detection: aqueous copper sulfate solution as spraying reagent (diethylamine first removed by warming) – brown spots of Cu-DDTC (replacement of central atom).
Sample preparation: Extraction from citric acid/phosphate buffer (McIlvain) pH 8 into chloroform.
Rf values: Pb, 0.00; Bi, 0.27; Cd, 0.34; Cu, 0.44; Hg, 0.56.

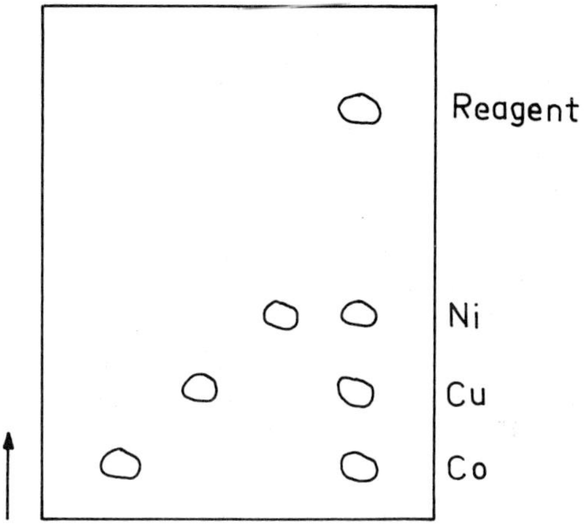

Figure 2

TLC separation of Co(III), Cu(II), and Ni(II) thiothenoyltrifluoroacetonates, according to *Müller* and *Rother* [294].
Stationary phase: silica gel (layer thickness 0.5 mm).
Mobile phase: chloroform/n-hexane (30:70, v/v).
Length of run: 10 cm. Duration of run: ca. 45 min.
Quantitative analysis: photometric determination after elution into microcuvettes.
Sample preparation: extraction with thiothenoyltrifluoroacetone in chloroform at pH 6.5 (noncomplexing buffer), reextraction of interfering excess of reagent with 0.01 M KOH.

1.1.2 Partition (TLC)

Support materials suitable for the polar, usually aqueous stationary phases used in partition chromatography of inorganic ions are alumina, silica gel, silanized silica gel, cellulose, and polyamides. Adsorption effects will clearly also play a role in many separations; in such cases, only the water content of the mobile phase, which leads to establishment of a partition system by adsorption of water on the support, will be regarded a criterion for distinguishing between adsorption and partition chromatography.

Table 3 presents just a modest selection of mobile phases and element groups to be separated. In general, mobile phase mixtures predominate which are composed of alcohols or ketones, as well as inorganic acids or their aqueous solutions. On the basis of these methods, separation processes have been developed for cations and anions which resemble those of classical qualitative analysis; they have been

Table 3

Partition chromatography of inorganic ions and of metal complexes on thin-layer plates. (In the case of less readily accessible original publications, the number of the table in *Brinkman*'s review [50] is given, marked with an asterisk.)

Specialized reviews:

Topic	Ref.
Precoated plates (various cellulose materials, polyamide); mobile phase, butanol/3N hydrochloric acid	[324]
Stationary phase alumina: separation of inorganic anions	[243]
Stationary phase polyamide: separation of metal ions	[244]

Mobile phase	Elements investigated	Ref.
Stationary phase: silica gel (selection)		
Amyl acetate/HCl (18:1)	Au, Hg, Pt, Pd, Rh	[433]
Acetone/HCl/ethyl acetate	Au, Ru, Pd, Rh, Os, Pt, Ir	*4.25
Acetone/HCl (99.5:0.5)	Co, Cu, Ni	[106]
Acetone/HCl	Li, Na, K, Rb, Cs, Be, Mg, Ca, Sr, Ba, Ti	*8.20
Butanol/benzene/N HNO_3/ N HCl (50:46:2.7:1.3)	Fe, Hg, Zn, Bi, Cd, Sn	*9.43/9.44
Acetone/methyl isobutyl ketone/ 6 N HCl/acetic acid	Fe, Cu, Co, Mn, Cr, Ni	[404]
Hexane/dioxane/methanol/butanone/ formic acid (40:20:5:5:1)	PAN complexes: Cu, Co, Fe	*5.64
Stationary phase: silanized silica gel		
Diisopropyl ether/ tetrahydrofuran/ 65% HNO_3	Rare earths	[167], [201], [375]
Stationary phase: alumina (selection)		
Alcohol/HCl mixtures	Re, Mo, V, W	*3.01
Butanol/12 N HCl (3:1)	Fe, Cu, Zn, Co, Cr, Mn, Ni	[108]
Acetone/acetylacetone/ 4 N HCl (91:4:5)	Cu, Co, Ni	[155]
Various mixtures with acetic acid	Acetylacetonates: Co, Mn, Ni, Cu, Cr, Fe	*5.58/ 5.59
Stationary phase: polyamide (selection)		
Butanol/acetone/HCl mixtures	Pb, Hg, Bi, Cu, Cd, Au, Sb, Fe, Te, Co, Mn, Ni, Cu, Zn, Sn, Pt, Pd, As, etc.	[244], [324]
Mainly methanol/HCl mixtures	Hg, Pb	*12.09
Butanol/3 N HCl	Review	[324]
Acetic acid/water/ methanol/propanol mixtures	Metal oxinates: Al, Zn, Zr, Be, La, Ba, Mg, Cd, Ca, Sr	[406]

Table 3 (Continued)

Mobile phase	Element investigated	Ref.
Stationary phase: cellulose (selection)		
Butanol/HCl/water	Cu, Co, Ni	[108]
Acetone/dioxane/HCl/water (50:25:13:12)	Al, Ti, V, Mn, Fe, Co, Ni, Zn, Zr, U	[121]
Acetone/butanol/6 N HCl (2:1:1)	Zn, Fe, U, V, Co, Mn, Al, Ti, Ni, Zr	[122]
Acetone/HCl/water	Ni, Mn, Co, Cu, Fe	*5.36
Various alcohols/6 N HCl	Zn, Fe, Cu, V, Co, Mn, Al, Ti, Ni, Zr	[322]
Butanol/HNO_3	Sc, Sm, Th, U	[179]
Dioxane/HCl, HNO_3	Sc, Y, La, Sm, Lu, Th, U	[305a]
HCl/water/methanol	Ba, Sr, Ca, Mg	[109]
HCl/water/dioxane	Be, Mg, Ca, Sr, Ba	[119]
12 N HCl/butanol (1:3)	Ag, Al, Cr, Ni, Mn, Pb, Co, Cu, Bi, Sn, Zn, Cd, Fe	[116]
Various mixtures with HCl, HNO_3, acetic acid, and water	Bi, Cd, Co, Cu, Fe, Mn, Ni Pb, U, V, Zn	[289] [290]
Various mixtures of butanol/HCl and other organic solvents	Cu, Pb, As, Bi, Cd, Sb, Sn, Hg	[373]
Acetone, methanol, butanol/ HCl/water mixtures	Al, Be, U, Fe, Cr, Ni, Mn, Co, Zn, Ba, Sr, Ca, Mg, Cd, Hg, Bi, Cu	*11.57 – 11.59
Various mixtures with ammonia	EDTA complexes: Cr, Mn, Fe, Co, Ni, Cu, Zn, Cd, Hg	[274]
Stationary phase: various gels		
Review of gel chromatography of inorganic polymers (polymeric anions), metal ions, simple anions, and of metal complexes on TLC plates [449].		
Chloroform, dioxane	8-Hydroxyquinolates: Cu, Zn, Al, Ga, In	[358]
Dioxane, benzene	β-Diketonates: Cr, Al, Co, Be	[357]

collected in the review by *Brinkman* [50] (see also Fig. 3).

Metal complexes have been successfully resolved by partition chromatography only in a few specific cases (e.g. as PAN chelates, acetylacetonates, EDTA complexes).

In recent years, considerable success has been achieved in the analysis of rare earths with silanized silica gel: all the elements belonging to this group can be completely separated with mobile phases of diisopropyl ether, tetrahydrofuran, and 65% nitric acid (see [375] and Fig. 4).

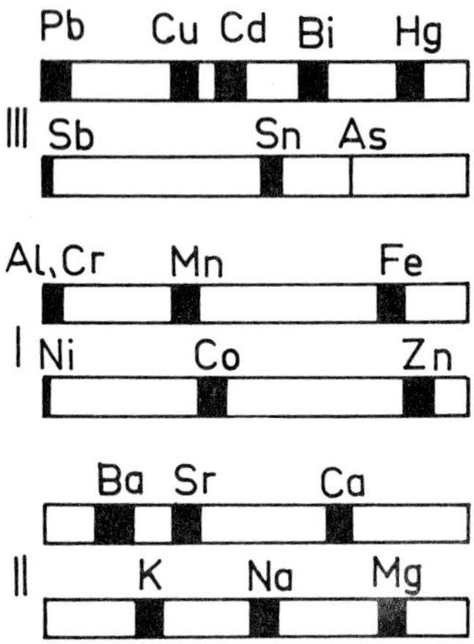

Figure 3

TLC group separation for qualitative analysis of cations, according to *Brinkman* [50], Table 12.20.
Stationary phase: silica gel (Merck) with 5% starch.
Mobile phases: I, acetone/3 N HCl (99:1, v/v); II, methanol/1-butanol/HCl (8:1:1, v/v); III, 1-butanol/benzene/1 N HNO_3/1 N HCl (50:46:2.7:1.3, v/v).
Length of run: 10 – 12 cm. Duration of run between 10 (I) and 45 min (II).
(The scheme is based on classical separation procedures.)

The numerous possible variations in the composition of the mobile phase frequently show selectivity effects and also the difficulty of a suitable choice for a specific analytical (separation) problem. Owing to the unclear nature of the separation effects, which preclude any statement of the predominating separation mechanism, these methods are less important than adsorption chromatography. However, they are often the method of choice for separation of ions without prior chelate formation.

Recent years have seen preliminary attempts to separate metal chelates with the aid of gel chromatography on TLC plates, a special kind of partition chromatography. Initial successes have been reported for the 8-hydroxyquinoline and β-diketone chelates with organic solvents as mobile phases (see Table 3).

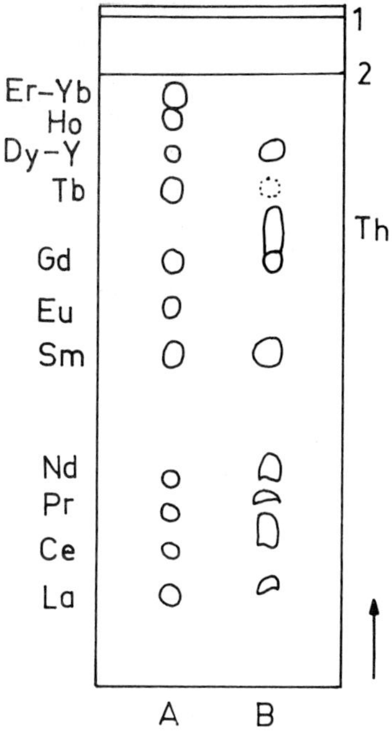

Figure 4

TLC separation of rare earths, according to *Jung* and *Specker* [201].
Stationary phase: silanized silica gel (Merck).
Mobile phase: diisopropyl ether/tetrahydrofuran/65% HNO_3 (100:70:6, v/v).
Duration of run: 60 min.
Detection: spraying with 1.5×10^{-2} M aqueous neothorin solution, then exposed to NH_3.
A: test mixture. B: analysis of Norwegian monazite.
1 and 2: 1st and 2nd solvent front.

The following Figures belonging to the Section "Partition Chromatography on TLC Plates" exemplify not only

- possibilities of group separation on silica gel (Fig. 3) and
- separations of rare earths on silanized silica gel (Fig. 4), but also

- a system containing polyamide for separation of cations (Fig. 5),
- a cellulose system for separation of alkaline earths (Fig. 6),
- the possibility of separating Cl-containing anions on silica gel (Fig. 7), as well as
- the separation of bismuth (and only partial separation of other cations) after streak application of the sample for photometric determination of this element (Fig. 8).

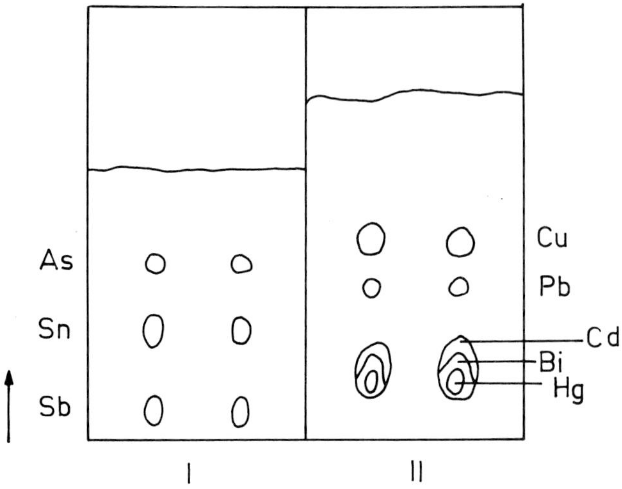

Figure 5

Cation separations on polyamide 11 layers, according to *Lederer* and *Rinalduzzi* [244].
Stationary phase: Polygram polyamide 11-precoated plates (Macherey-Nagel, Düren, FRG).
Mobile phases: I, butanol/1 N HCl (sat.); II, butanol/4 N HCl.
Detection: spraying with hydrogen sulfide solution or ammoniacal/ethanolic 8-hydroxyquinoline solution.
Samples: solutions of cations in 1 – 3 N HCl.

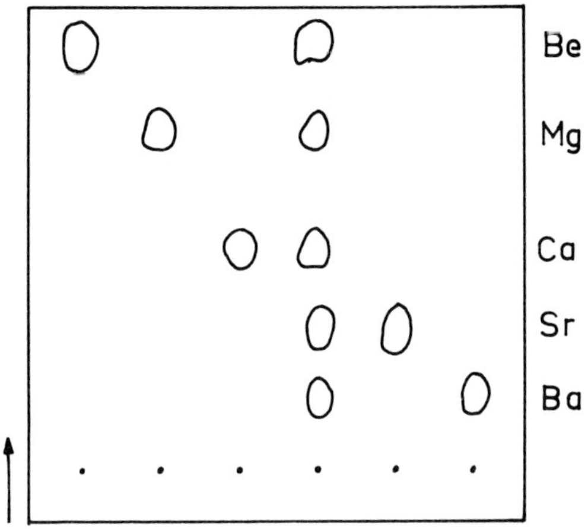

Figure 6

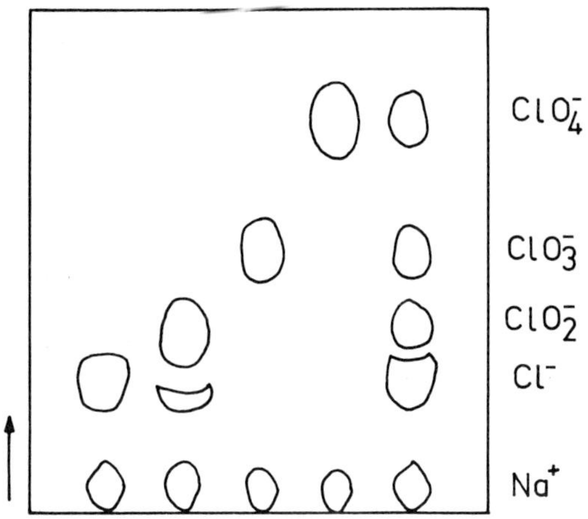

Figure 7

Figure 6

TLC separation of alkaline earths on cellulose, according to *Gagliardi* and *Likussar* [119].
Stationary phase: MN cellulose powder 300 HR (Macherey-Nagel, Düren, FRG), mean particle size 10 μm.
Length of run: 10 cm.
Mobile phase: methanol/HCl/water (73:12:15, v/v).
Detection: spraying with NH_3 and 2% ethanolic 8-hydroxyquinoline solution.
Rf values: Be, 0.95; Mg, 0.77; Ca, 0.50; Sr, 0.30; Ba, 0.12.

Figure 7

TLC separation of Cl-containing anions on silica gel, according to *Seiler* and *Seiler* [345].
Stationary phase: silica gel MN – S – HR (Macherey-Nagel, Düren, FRG).
Length of run: 10 cm. Duration of run: 70 min.
Mobile phase: isopropanol/tetrahydrofuran/NH_3 (50:30:20, v/v).
Detection: spraying with 0.1% solution of methyl red in 50% ethanol.
Quantitative analysis: elution, measurement of β-radiation emitted by the salts irradiated prior to TLC.
Samples: 1 μl each of 0.5 M solutions.

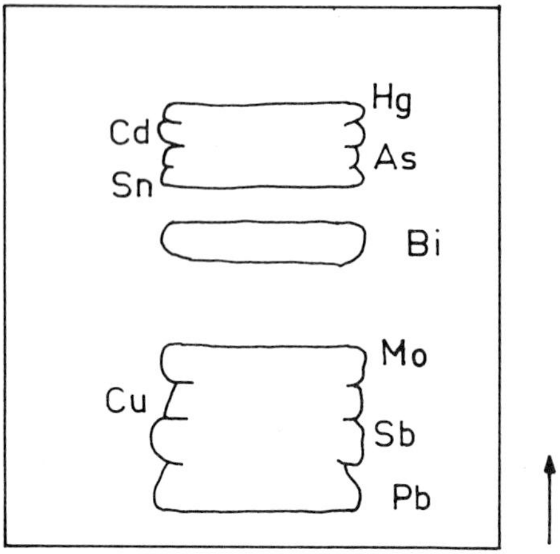

Figure 8

TLC separation of bismuth from other elements of the H_2S group, according *Gagliardi* and *Likussar* [120].
Stationary phase: cellulose.
Mobile phase: tert-butanol/HCl/H_2O (76:8:16, v/v).
Detection: with 0.2% potassium iodide solution.
Quantitative analysis: elution, extraction at pH 2–7 with morpholinium N-morpholinecarbodithioate in chloroform, photometric determination at 365 nm.

1.1.3 Reversed Phases (TLC)

Table 4
Reviews of reversed phase (RP) TLC

Topic	Year	Ref.
Survey of RP partition chromatography in inorganic chemistry	1970	[67]
Theory of the cellulose/tributyl phosphate system	1970	[16]
Theory of the chromatographic process	1971	[18]
Separation and identification of metal ions	1971	[76]
Separation of inorganic ions	1966	[215]
Separation and enrichment	1972	[7]
RP with liquid anion exchangers	1967	[49]
	1972	[46]
RP with alkylamines and quaternary ammonium salts	1967	[295]
RP with tri-n-alkylarsine and tri-n-alkylarsine oxide	1971	[246]
RP with bromide-containing eluents	1972	[51]
RP with HNO_3-containing eluents	1973	[247]

Table 5

Reversed phase chromatography on thin-layer plates.

(In the case of less readily accessible original publications, the number of the Table in *Brinkman*'s review [50] is given, marked with an asterisk.)

Impregnating agent (Support material)	Elements investigated	Reference
Tri-n-butyl phosphate (SiO_2)	Rare earths	*7.01
(Polyurethane foam)	Au (separated from Zn, Co, Ni, Fe, Sb, U, Bi, Pd)	[42]
(Polyurethane foam)	Fe separated from Co, Ni	[45]
(Cellulose)	Cd, Co, Mn, Pd, Zn	[16]
(Cellulose)	Rare earths	[69]
(Cellulose, silica gel)	Ti, V, Cr, Mn, Co, Ni, Cu, Zn, Mo, Ru, Rh, Pd, Ag, Cd, In, Sn	[17]
(Teflon)	Re, Mo	[308]
Hydrogen di(2-ethylhexyl)phosphoric acid (Cellulose, silica gel)	Rare earths	*7.01, 7.05, [168]
(Cellulose)	Alkaline earths	[68]
Dibutylphosphoric acid (Silica gel)	Pr, Nd, Sm, Eu, Gd, Tb, Dy, Ho, Er	*7.01
Tetrabutylhypophosphoric acid (Teflon)	Np, Pu, Am, U, Fe, Ni	*7.12
Tri-n-octylphosphine oxide (Cellulose)	Co, Cu, Mn, Ni	*5.01
(Cellulose)	Co, Cu, Mn, Ni, Zn, Re, Te Pb, Sb, Se, Th, Be, Ca, Fe	*11.03
Methylene bis(di-n-hexylphosphine oxide) (Cellulose)	Zr, Nb	[203]
Methylene bis(di-2-ethylhexyl)phosphine oxides (Cellulose)	Rare earths	[310]
Tri-n-octylarsine oxide	Systematic survey	[46]
High-molecular amines	Systematic surveys	[47, 48, 52]
	Mn, Ni, Co, Cu, Cd, Ag, Zn, Pb	[198]
	Rare earths, Th, U, Zr	[360, 361]
(Silica gel)	Cu, Co, Mn, Zn, Fe	[49]
(Cellulose)	Bi, Cd, Co, Cu, Pb, Mn, Zn	*9.03
Polyethylene imine (Cellulose)	Polyphosphates	[408]
Chelating agents		
Dithizone (SiO_2)	Hg, Cu, Cd, Ni, Zn	[84, 376]
8-Hydroxyquinoline (Silica gel)	Cu, Fe, Ni, Co	[298]
Thiourea (SiO_2)	Se, Te	*2.07/2.08
Fe, Pb diethyldithiocarbamate (Cellulose)	Ag, Cu, Pb, Cd	[6]

Table 5 (Continued)

Impregnating agent (Support material)	Elements investigated	Reference
Phenylbenzohydroxamic acid (Cellulose)	35 Metal ions	[115]
Tetrasubstituted pyrazole	Numerous metal ions	[53]
Acetophenone (Polyethylene)	Zr, Hf	[74]
Thenoyltrifluoroacetone	Fe, Co, Zn, Sr, Zr, Cs, Ce	[238]
(Cellulose)	Ni, Mn, Co, Cu, Zn, Fe	[82]
Monothiothenoyltrifluoroacetone (Cellulose)	Cu, Ni	[169]
p-Toluidine	Co, Zn, Ni, Ti, Th, Fe, Zr, Cu, As, Pb, Cd, U, V, Mo	[377]
Nitrilotriacetic acid	Various metal ions	[378]

One special kind of partition chromatography employs impregnated layer materials or papers. The impregnation agents are inorganic or organic compounds which can react with the ions to be separated, usually forming complexes or compounds that are sparingly soluble in the mobile phase.

A prerequisite for the feasibility of this chromatographic method is that the impregnation agent be insoluble in the mobile phase, which is usually an aqueous solution of inorganic acids or a mixture of organic solvents (acetone, alcohols) with organic or inorganic acids.

In contrast to the situation in the adsorption chromatography of metal chelates, in this variant the chelating agent constitutes the stationary phase and is not already bonded to the metal ion before the chromatographic separation process begins. Moreover, the mobile phase is polar whereas nonpolar eluents are used in adsorption methods. This is why the technique is called reversed phase chromatography (as opposed to adsorption chromatography). It has meanwhile become very important in HPLC, thanks to the availability of special stationary phases (see pertinent literature in Table 1).

Apart from the chelating agents already known from adsorption chromatography (see Table 5), other impregnating agents include alkylphosphates and alkyl-phosphine oxides and alkylarsine oxides, as well as high-molecular amines and their quaternary salts. These methods enjoy widespread use in layer chromatography for a wide range of very difficult separation problems. Several surveys of the field have therefore been published (see Table 24).

On use of high-molecular amines and alkyl phosphates as impregnating agents (= stationary phases), ion exchange processes come to the fore: these systems represent a transition to ion exchange chromatography which is presented in Section 1.1.4. However, they are usually still classified as belonging to reversed phase chromatography.

1.1.4 Ion-Exchange (TLC)

Apart from liquid ion exchangers (review by *Green* [141]) used to soak cellulose or silica gel, which have already been mentioned in the Section on reversed phase chromatography (Section 1.1.3), there are various other materials that can be used for ion-exchange chromatography on layers:

1) Modified celluloses (e.g. with 2-diethylaminoethyl (DEAE) or carboxymethyl (CM) groups: $R-O-CH_2CH_2-N(C_2H_5)_2$/strongly basic anion exchangers; $R-O-CH_2-COOH$/weakly acidic cation exchangers).
2) Celluloses with chelating groups (chemically bonded by azo coupling reactions).
3) Anion and cation exchangers based on synthetic polymers (divinylbenzene-styrene).
4) Chelate exchangers (Chelex 100: iminodiacetate groups bonded to divinylbenzene-styrene).
5) Inorganic materials (like phosphates) with ion-exchange properties.

Materials belonging to these five groups have been successfully used for TLC separations, mainly of cations. Mixtures of different exchangers have also been used for specific separation problems (see ref. [50]). However, these methods are only of minor importance in quantitative inorganic analysis: detection and quantitative determination depend upon prior spraying with color reagents and evaluation of the plates; the ready elution of chelates as in adsorption chromatography on layers cannot be applied for subsequent photometric determination.

Table 6

Ion-exchange chromatography on thin-layer plates (selected examples).
(In the case of less readily accessible original publications, the number of the table in *Brinkman*'s review is given, marked with an asterisk.)

a) *Basic 2-diethylaminoethyl (DEAE) cellulose*

"Mobile phase" system	Elements investigated	Ref.
0.4 M ammonium sulfate	Cr, Fe, Cu, Zn, Ni	*5.20
Acetone-HCl	Fe, Al, Cr, Co, Ni, Zn, Mn, Pb, Hg, Bi, Ca, Cu	
(Survey of various ion-exchange celluloses)		*9.42
Alcohols-HNO_3	Numerous elements	[178]
Alcohols-HCl	Numerous elements	[236]
Butanol-3 N HCl	Numerous elements	[324]
Oxalic acid-HCl	Numerous elements	[237]
Methanol-HSCN	Numerous elements	[307]

b) *Carboxymethyl (CM) cellulose*

"Mobile phase" system	Elements investigated	Ref.
0.4 M ammonium sulfate	Cr, Fe, Cu, Zn, Ni	*5.20
Acetone-HCl	Fe, Al, Cr, Co, Ni, Zn, Mn, Pb, Hg, Bi, Ca, Cu	*9.42
Various buffer solutions	Multiply charged ions	[248]
Salt solutions (NaCl, $NaClO_4$, KCl/HCl)	Numerous elements	[79]

c) *Sulfoethyl (SE) cellulose*

"Mobile phase" system	Elements investigated	Ref.
$H_2SO_4/(NH_4)_2SO_4$	Inorganic ions	[363]

d) *Cellulose phosphate*

"Mobile phase" system	Elements investigated	Ref.
$H_2SO_4/(NH_4)_2SO_4$	Numerous metal ions	[362]

e) *Celluloses with chelating groups (chemically bonded)*

Chelating agent	Elements investigated	Ref.
$-N-C_6H_4-N=N-R$	Cu, Th, Hf, U, etc.	*11.90
Dithizone, oxine (bonded by diazo coupling to CM-cellulose)	Pb, Zn, Cd, Mn	[21]

Table 6 (Continued)

f) *Inorganic exchangers*

Stationary phase	Elements investigated	Ref.
Tin arsenate	57 Metal ions	[177]
Zinc antimonate	49 Metal ions	[327]
Titanium arsenate	47 Metal ions	[326]
Thorium tungstate	Hg from other metal ions	[83]
Zr, Ti phosphate (on SiO_2)	Li, Na, K, Cs, Tl, Ca, Ba, Sr	[4]
Zr phosphate	Ni, Co, Pb, Fe, Ag, Hg, Cd, Cu	[219, 452]
Ce(IV) phosphate	Numerous elements	[220]
NH_4 phosphorododecamolybdate	Na, K, Rb, Cs, Ba, La, Sr, Y	[253]
Heteropolyacids: P-, As-, Ge dodecamolybdates	Na, K, Rb, Cs	[257]
Zn-, Co hexacyanoferrate(II)	Na, K, Rb, Cs	[253]
Magnesium fluoride	Cr, Mn, Fe, Ni, Cu, Zn, Al, Co, Cd, Hg, Pb	*9.23

g) *Organic (polymeric) ion exchangers*

Name of exchanger	Elements investigated	Ref.
Cation exchangers:		
Dowex 50-X4 (comparison with CM cellulose)	Numerous elements	[248, 249]
Amberlite CG-120, -400	Inorganic ions	[103]
Dowex 50W-X8, 1-X8	U, V, Fe, Co, Ni, Cu, Ag, Au, Cd, Bi	[356]
Dowex 50W-X8 (K^+ form,	Anions	*16.05
+ silica gel)	Polythionates	[372]
Anion exchangers:		
Amberlite CG-120, -400	Various metal chloro complexes	[103]
Dowex 2-X8 (Cl form), Ionex 25-SB (acetate form)	Phosphates	[232]
Chelate exchangers:		
Chelex 100	Hg, Ag, Pb, Sb, Bi, Cd, Cu, Cr, Fe, Ni, Al, Zn, Co, Mn	*12,01 – 12.08, [99b]

In contrast, ion exchange chromatography on columns has acquired far greater importance in the separation and especially in the enrichment of ions (see Sections I/1.2.4 and III/3).

Table 6 illustrates the role of this specialized separation method for ionic inorganic substances by means of a selection of separation systems and elements investigated.

1.2 Column Chromatography (CC)

Liquid chromatography on columns is far less widespread than thin-layer chromatography in inorganic analysis. Up to the early 1970's it was used primarily as ion exchange chromatography for separation, preliminary fractionation, or enrichment of trace elements from large volumes of aqueous solutions. The reasons for the comparatively scant utilization of column chromatography lie mainly in the better and more versatile methods of detection and quantitative determination available with thin-layer chromatography (see Section II/1).

During the past ten years, though, rapid developments in the production of high-performance support materials and the development of sensitive detectors, which can be coupled directly with the separation column, have focused interest on column chromatography in the form of high performance liquid chromatography (HPLC) as a tool for quantitative inorganic analysis, of special use in trace analysis.

While there is a large and increasing number of publications about HPLC of organic substance, less than one hundred papers have so far been published on the application of this technique to inorganic analysis (see *Schwedt* 1979 [397]: review article). Less than half of these papers are concerned with quantitative, and very few with practical applications.

The reason lies mainly in the fact that inorganic analysis normally has no contact with this instrumental method, because it is still regarded as belonging to the domain of organic analysis. Preliminary results reported in recent years regarding the separation performance of high pressure liquid chromatographic systems in the case of metal chelates with particle sizes between 40 and 5 μm hold promise of successful general application in this field, particularly since TLC separations can often be transferred to separations in columns (see Section 1.2.1).

Extension of the horizon of chromatography beyond that of adsorption, liquid-liquid distribution, and ion-exchange chromatography has already come from reversed phase and partition chromatography with chemically bonded phases, whose development is still continuing. HPLC is of great importance for trace analysis, especially in view of the increasing possibility of coupling with element-selective and sensitive detectors (see Section II/2).

The following Sections deal almost exclusively with publications from the field of modern column chromatography with on-line coupling of a detector. No distinction is made between "forced-flow" and "high pressure (performance)" chromatography, which reflect little more than the particle size of the support material (about 40 and 5 – 10 μm, respectively).

1.2.1 Adsorption Chromatography (CC)

Work published so far on adsorption chromatography of metal complexes by modern HPLC can be subdivided into (i) separations of particular, frequently isomeric, chelates of one or a few elements (Table 7) and (ii) separations of several elements in the form of known complex compounds.

Transfer of TLC separations to column chromatography is illustrated, for example, by the adsorption chromatography of metal benzylmethyldithiocarbonates (Fig. 9).

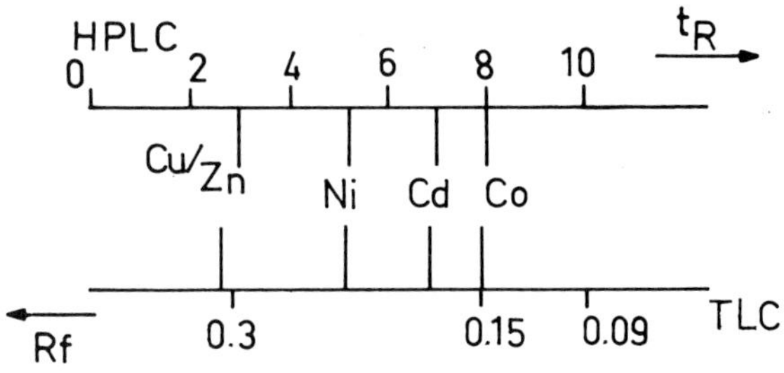

Figure 9

Comparison between TLC and HPLC separations. Adsorption chromatography of metal benzylmethyldithiocarbamates, according to *Heizmann* and *Ballschmiter* [162].
TLC: stationary phase, silica gel
mobile phase, benzene/cyclohexane (3:1, v/v).
HPLC: stationary phase, LiChrosorb® Si 60, particle size 40 μm (column dimensions, 350 × 2 mm)
mobile phase: benzene/cyclohexene (t_R in min).

Table 7
Adsorption chromatography of metal complexes by HPLC.

Element studied	Compounds	Ref.
Co	Metallocarboranes	[95]
Cr	Various keto complexes	[195]
Fe	Ferrocenes	[159]
Hg	Organomercurials	[117]
Ni	Dialkyldithiocarbamates	[257, 258]
Fe and other transition metals	Cyclopentadienyl complexes	[158]
Ir, Rh	Triphenyl phosphinates	[93]
Cu, Ni	N,N'-Ethylenebis-(acetylacetone imine)-, N,N'-ethylenebis(salicyl-aldimine) chelates	[422]
Cu, Ni, Co	1,2-Diketobisthio-benzhydrazones, dialkyl-dithiocarbamates, 1,2-di-ketobisthiosemicarb-azones	[162]
Hg, Cu, Pb, Zn	Diacetylbis(thio-benzhydrazones)	[161]
Pb, Zn, Cd, Hg, Cu, Ni, Co	Dithizonates	[311, 261]
Co, Ni, Cu	Diethyldithiocarbamates	[423]
Cu, Co, Zn, Pb	N-Disubstituted dithio-carbamates	[250]
Hg, Cd, Pb, Cr, Bi, Cu	Diethyldithiocarbamates	[293]
Zn, Cu, Mn, Ni, Pb, Cr, Co, Cd, Fe	Diethyldithiocarbamates	[259, 311]
Fe, Co, Ni	8-Mercaptoquinolates	[223a]
Ni, Pd, Pt, Rh, Ir	1-Hydroxy-2-pyridinethio-nates	[223a]

As in TLC (see Section 1.1.1), the possibility of separating dithiocarbamates and dithizonates by HPLC has so far been studied most thoroughly. However, application to practical analytical problems has yet to come. Various silica gels having particle diameters of 5 – 10 μm or 30 – 40 μm have been used as adsorbents. A comparison of separation performance as a function of particle diameter is shown in Figure 10. Depending upon the separating material and the na-

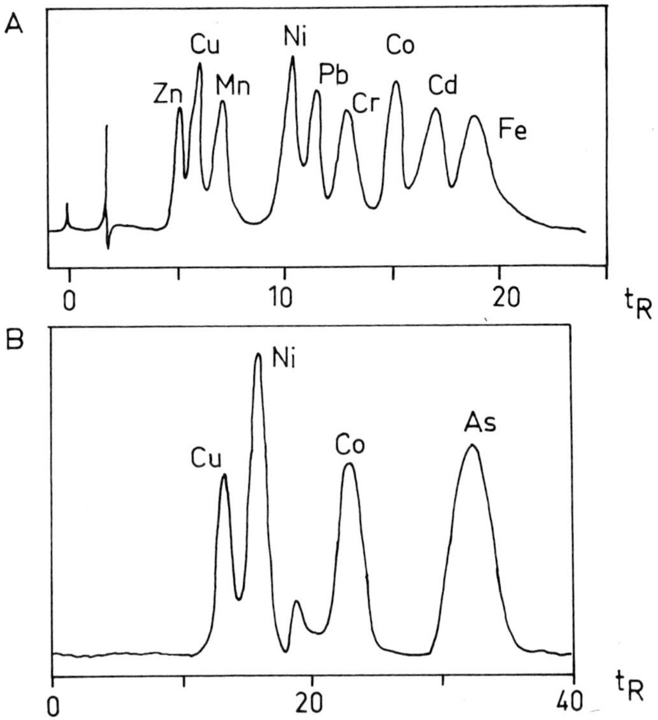

Figure 10

Separation of metal dithiocarbamates by adsorption chromatography on silica gels of various particle sizes. Comparison of separation performance.

A: Separation of metal diethyldithiocarbamates according to *Liska* et al. [259].
Stationary phase: LiChrosorb® Si 60, particle size 10 μm (column dimensions 200 × 4.6 mm).
Mobile phase: 10% chloroform in cyclohexane.
Linear velocity: 0.13 cm s^{-1} (t_R in min).
Pressure: 2.5 MPa.
Number of theoretical plates: ca. 7000 m^{-1}.
Detection: UV at 254 nm.
Sample: mixture containing 5.6 × 10^{-7} mol/dm^3 of each compound.

B: Separation of metal diethoxyethyldithiocarbamates, according to *Heizmann* and *Ballschmiter* [162].
Stationary phase: LiChrosorb® Si 60, 40 μm (column dimensions 350 × 2 mm).
Mobile phase: 4% acetonitrile in carbon tetrachloride.
Flow rate: 8 cm^3 h^{-1} (12 min), then 20 cm^3 h^{-1}.
Number of theoretical plates: ca. 1300 m^{-1}.
Detection: at 360 nm.

ture of the chelates up to 10 elements can be separated from one another in less than 30 min. (Fig. 10A).

HPLC techniques also permit use of programmed gradient elution, leading to a cut in analysis time and improvement of separation (e.g. Fig. 11). Gradient techniques are necessary in particular whenever some substances exhibit very low Rf values in TLC-optimized separations of metal chelates, and therefore have very long retention times in the column.

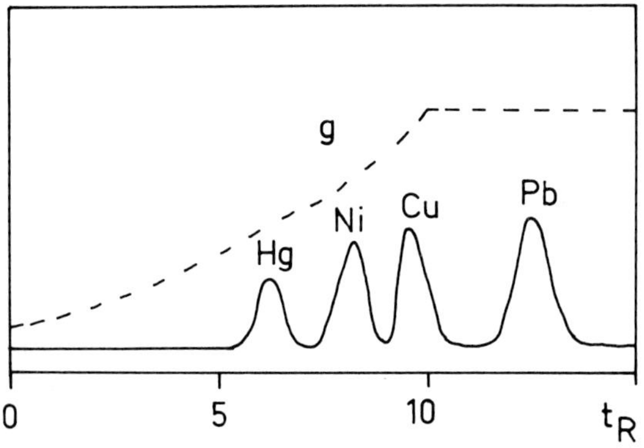

Figure 11

Adsorption chromatography of metal diacetylbisthiobenzhydrazone chelates, according to *Heizmann* and *Ballschmiter* [161].
Stationary phase: LiChrosorb® Si 60, 30 μm (column dimensions 500 × 2 mm).
Mobile phase: n-heptane/benzene (g = gradient elution – 40% benzene in n-heptane to 100% benzene).
Flow rate: 1.5 cm^3 min^{-1} (t_R in min).
Detection: UV at 360 nm.
Sample: 50 μm (containing 10^{-9} g of each metal).

1.2.2 Partition Chromatography (CC)

Partition chromatography (between two liquids) has hitherto played only a very modest role in the HPLC analysis of inorganic compounds. A comparatively large number of elements have been successfully separated, e.g. as acetylacetonates, in a ternary phase system (see Fig. 12) or, in the case of rare earths, as nitrates on silanized silica gel [439, 375].

Hardly any results are available regarding the applicability of chemically bonded phases: a NH_2 phase has been used, however, for separation of β-keto amine complexes [117a].

Table 8
Partition chromatography by HPLC.

Element/Compound	Stationary phase	Ref.
Cr/Isomeric arenetri-carbonyl complexes	Carbowax	[142, 430]
Cr, Fe/Metallocene	Carbowax	[91]
Be, Cu, Fe, Al, Cr, Ru, Zn, Co, Zr/Acetyl-acetonates	Ternary phase system	[175]
Rare earths/Nitrates	Silanized silica gel	[439, 375]
Transplutonium elements/Salts	Di(2-ethylhexyl)ortho-phosphoric acid	[62]
Co, Ni, Cu, Pd/ Chelates with tetra-dentate β-keto amines	3-Aminopropyltrieth-oxysilane (chemically bonded to silica gel)	[117a]

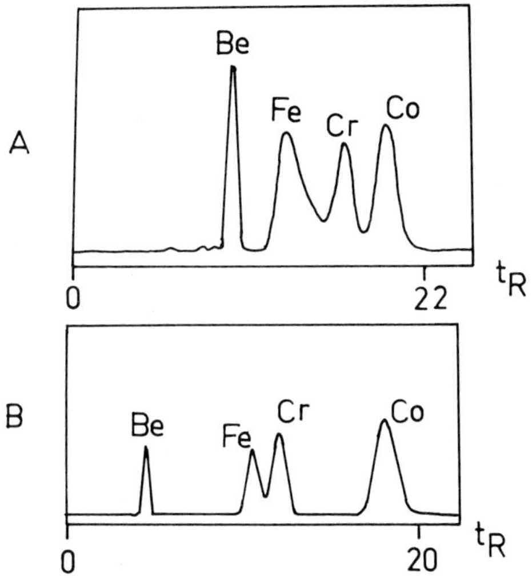

Figure 12

Partition chromatography of metal acetylacetonates, according to *Huber, Kraak*, and *Veening* [175].
Ternary phase system: water/ethanol/2,2,4-trimethylpentane.
Stationary phase: 0.343 : 0.641 : 0.016 (parts by mass) on kieselguhr.
Mobile phase: 0.0007 : 0.022 : 0.977 (parts by mass).
Column: A, 100 × 2.7 mm; B, 500 × 2.7 mm.
Linear velocity: A, 0.29 mm s^{-1}; B, 2.6 mm s^{-1}.
Detection: UV at 310 nm.
A, without ligand in mobile phase; B, with low concentration of acetylacetone in mobile phase.

1.2.3 Reversed Phase Chromatography (CC)

Table 9

Liquid chromatography with reversed phases (reversed phase HPLC).

Compound	Element investigated	Ref.
Organometallic cpds.	As, Pb, Hg, Sn	[54]
Organometallic cpds.	Fe	[137]
Dibenzo-18-crown-6 complexes	Hg	[269]
Carbonyl-, triphenylphosphine complexes	Fe, Co, Cu, Cr	[94]
β-Diketonates, β-keto amine chelates	Cu, Co, Cr, Al, Be, Ru, Rh, Fe, Ni	[415]
Chelates with Schiff bases	Ni, Cu, Pd	[422]
Diethyldithiocarbamates	Se, Pb, Ni, Cu (survey)	[392] [393, 395]
Tetramethylenedithiocarbamates	Cd, Cr, Ni, Co, Pb, Cu, Hg, Zn, etc.	[395]
Tetradentate β-keto amine chelates	Co, Ni, Cu, Pd	[117a]
Piazselenoles	Se	[392, 399] and [440]
Reversed phase paired ion chromatography		
Tri(2,2′-bispyridyl)Ru(II)- + n-heptane-sulfonate	Ru	[426]
Inorganic anions + cetyltrimethyl-ammonium salts		[335]

Chemically modified silica gels, containing C_2 to C_{18} alkyl groups bonded via silicon atoms, exhibit pronounced chemical stability and permit highly reproducible separations of polar to nonpolar substances with mobile phases made up of water and relatively polar organic solvents such as methanol or acetonitrile.

Successful separations of chelates with Schiff bases as ligands and of dithiocarbamates (see Table 9) have been reported. Although only few publications exist in this very new field of application to inorganic substances (new as regards chemically bonded phases as opposed to reversed phase systems in TLC), the inherent performance of these systems will certainly lead to further applications. It is essential, however, that the metal chelates be relatively stable to water and have a significant solubility in the mobile phase.

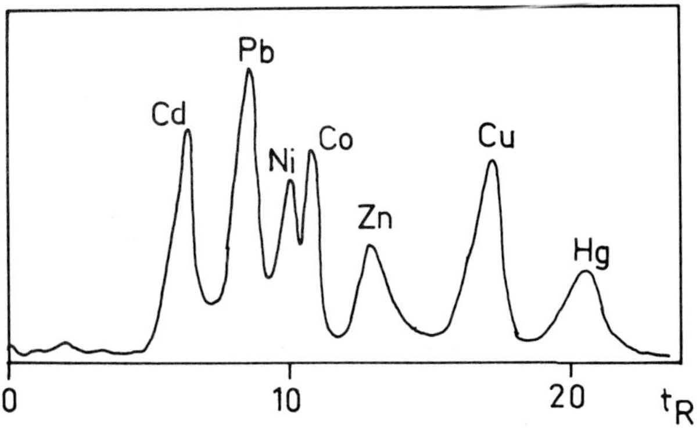

Figure 13

Reversed phase HPLC of metal tetramethylendithiocarbamates, according to *Schwedt* [395].
Stationary phase: LiChrosorb® RP-8, particle diameter 8 µm (column dimensions: 250 × 4.6 mm).
Mobile phase: methanol/water (70:30, v/v).
Flow rate: 1.2 $cm^3\ min^{-1}$ (t_R in min).
Pressure: 70 bar.
Detection: UV at 254 nm.

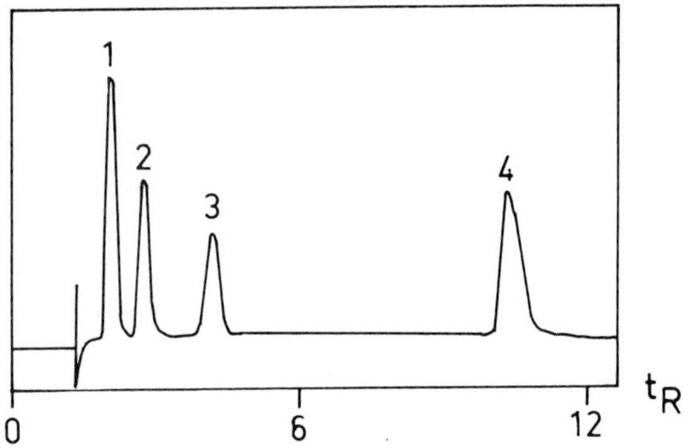

Figure 14

Reversed phase chromatography of ion pairs, according to *Reeve* [335].
1, iodate; 2, bromate; 3, bromide; 4, iodide.
Stationary phase: SIL 60-D-10-CN (Chrompack), chemically bonded phase with nitrile groups (column dimensions 250 × 4.6 mm).
Mobile phase: methanol/aqueous solution of 0.1 mol/dm^3 Na_2HPO_4, 0.1 mol/dm^3 KH_2PO_4, and 0.1% (w/v) cetyltrimethylammonium bromide (40:60).
Flow rate: 2 $cm^3\ min^{-1}$.
Sample volume: 10 µm.
Detection: UV at 215 nm.

Reversed phase chromatography is unsuitable, e.g. for separation of metal β-diketonates (see ref. [415]); adsorption and partition chromatography (with ternary phases) proves superior in this case.

Reversed phase chromatography of ion pairs with the cetyltrimethylammonium ion as counterion has recently been described for the first time (see Fig. 14) as a method for separating a number of anions. Analysis times of less than 10 min for separation of several anions and the possibility of detection in the UV region are features of this new method. It is still too early to generalize about the significance of this chromatographic method for inorganic analysis (it has frequently been used for organic substances).

1.2.4 Ion Exchange/Ion Chromatography (CC)

In contrast to reversed phase chromatography, ion-exchange chromatography on pressure-stable materials of small particle size has already conquered a firm place in inorganic trace analysis. Numerous elements can be separated by gradient elution or also on isocratic operation, and can be sensitively determined in the fractionated eluate or in a flow-through cell after post-column derivatization with a metal reagent (see Section II/2.3).

Apart from organic polymeric ion exchangers, Zeokarb materials and zirconium phosphate have been used, as has the liquid ion exchanger di-(2-ethylhexyl)orthophosphoric acid (see Table 10).

New groups of ion exchangers may have a silica gel matrix, to which 8-hydroxyquinoline molecules or amide groups are chemically bonded, a cellulose matrix, to which various ligands are likewise linked via chemical bonds, or belong to the class of crown ether ion exchangers.

Materials with chemically bonded selective ligands are used primarily for enrichment and separation of individual elements from aqueous solutions.

The use of complexing exchangers with crown compounds (cyclic polyethers) as anchor groups has various characteristic features (see publications from *Blasius*' group). These materials are not sparingly soluble polyelectrolytes, but uncharged macromolecules which exchange cations bound to anions, i.e. salts, and form polyether complexes. Binding is

Table 10

Separations of metal ions and anions, as well as ionic complexes by means of ion exchangers (selection).

Exchanger	Elements investigated/Complex	Ref.
Zeokarb	Halides	[104]
Zr phosphate	Alkali metals, alkaline earth metals	[10, 11]
Di(2-ethylhexyl)orthophosphoric acid	Transplutonium elements	[62]
Organic polymers:	Aminocarboxylic acid chelates: Cu	[196]
	NTA-, EDTA complexes: Cu	[197]
	Na, K, Rb, Cs	[174]
	Alkali metals, alkaline earth metals	[132]
	As, Bi, Sb – Ni, Pd, Pt – Pb, Cu, Fe, Hg, Sn	[352]
	Cd, Zn, Fe, Pb, Cu, Co, Mn	[212, 213]
	Zn, Pb, Cu, Mn, Ni, Mg, Ca, La, Th, Zr	[113]
	Cu, Co, Mn, V, Ni	[210]
	Cd, Zn, Fe, Pb, Cu, U, Co, Mn, V, Ni	[211]
	Alkali metals	[180]
	Lanthanides	[92]
	Rare earths	[63]
	Nitrate/nitrite	[81]
	Polythionates	[442, 443]
	17 cations and anions	[367]
	Condensed phosphates	[444]
Chelate ion exchangers with thioglycolic acid as functional group:	Pb, Bi, Hg, Ag, Sn, Sb, Au, Cd, Cu, As	[319]
8-Hydroxyquinoline immobilized an silica gel	13 metal ions	[182, 184]
Amide group-containing ion exchangers	Pt elements, Au, Th, U, Zr	[321]
Crown ether ion exchangers	Cations and anions	[30 – 33]
Celluloses with chemically bonded ligands: 1-(2-Hydroxyphenylazo)-2-naphthol (Hypha)	Fe, Cu, Zn, Pb	[58]
4-(Pyridyl-2-azo)resorcinol (PAR)	Fe, Cu, U	[59, 60]
Bis(salicylaldehyde)-ethylenediamine (Salen)	Fe, Cu, U	[59, 60]

associated with a extensive stripping of the solvation sheath; the stability of the polyether complexes depends upon the cation, the corresponding anion, the solvent (= the mobile phase), and the size of the polyether ring. Since the pure solvent is simultaneously used as eluent, no impurities reach the eluate. Work published to date demonstrates the suitability of these exchangers for trace and microanalysis.

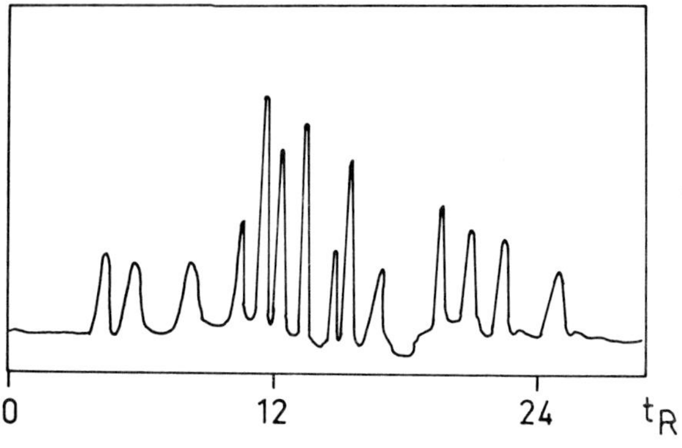

Figure 15

Ion exchange chromatography of lanthanides, according to *Elchuck* [92].
Sequence of separated lanthanides:
Lu, Yb, Tm, Er, Ho, Dy, Tb, Gd, Eu, Sm, Nd, Pr, Ce, La.
Stationary phase: Aminex A-5, 13 μm particle size* (column dimensions: 100 × 4 mm).
Mobile phase: linear program from 0.17 mol/dm^3 hydroxybutyric acid to 1.0 mol/dm^3 over 20 min, pH 4.6.
Flow rate: 0.8 cm^3 min^{-1} (t_R in min).
Detection: spectrophotometry at 600 nm after reaction with arsenazo I.

Ion chromatographic methods in columns are performed under normal pressure or slightly elevated pressure, or as HPLC (see Fig. 15 for HPLC ion exchange chromatography). No hard and fast boundary can be drawn to HPLC. All methods which make use of direct coupling between detector and separation column are generally classed as modern liquid chromatography (HPLC) regardless of operating pressure.

A variant of classical ion exchange chromatography is termed ion chromatography. It is distinguished above all by continuous conductometric detection and use of two columns. The interfering electrolyte

* Separations of comparable efficiency have also been accomplished on chemically bonded exchange materials, such as Nucleosil SCX and Partisil-10 SCX.

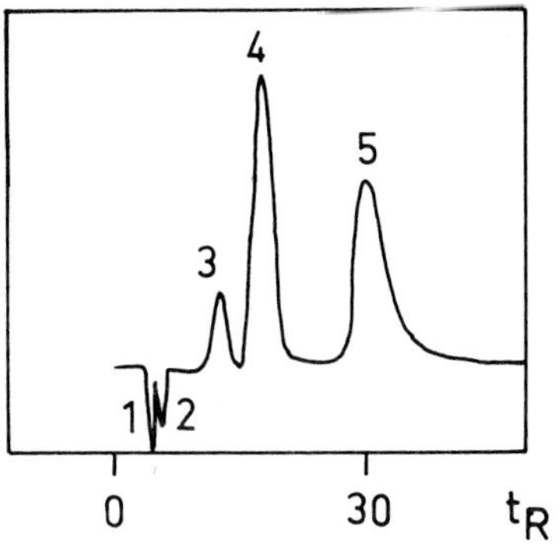

Figure 16

Ion exchange chromatography of anions, according to *Franks* [104].

1, nitrate/fluoride; 2, sulfate; 3, chloride; 4, bromide; 5, iodide.

Stationary phase: Zeokarb 225 (cadmium form).

Mobile phase: 0.0025 mol/dm^3 cadmium acetate in water.

Flow rate: 0.4 $cm^3\ min^{-1}$ (t_R in min).

Sample volume: 50 µl with 1000 ppm each of sulfate, nitrate, fluoride, and 1 ppm each of halide ions.

Detection: potentiometry (silver/silver chloride microelectrode), flow-through cell.

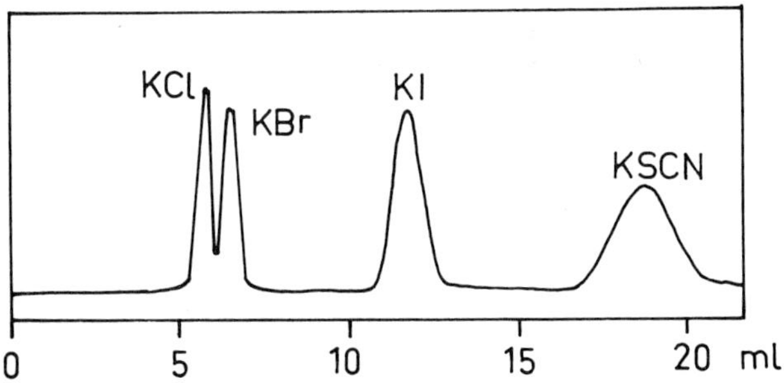

Figure 17

Separation of various potassium salts on a complexing exchanger with cyclic polyether anchor group, according to *Blasius et al.* [32].

Stationary phase: exchanger produced by condensation with dibenzo[21]crown-7 anchor groups (details, see ref. [32]), column dimensions 300 × 4 mm.

Mobile phase: water.

Flow rate: 5 $cm^3\ h^{-1}$.

Pressure: 3 – 15 bar.

Detector: differential refractometer.

Sample: 0.42 mg KCl; 0.60 mg KBr; 1.60 mg KI; 1.77 mg KSCN.

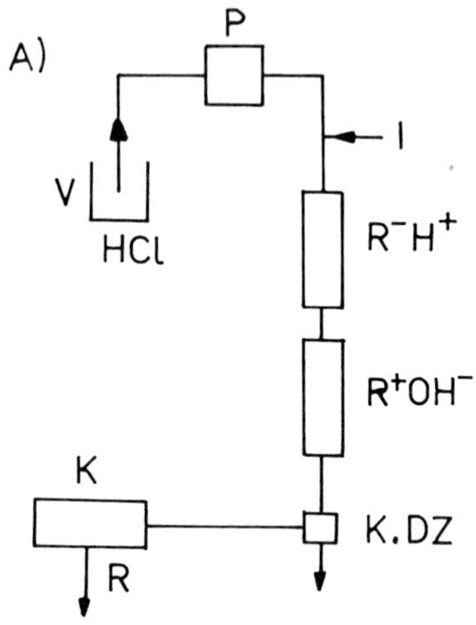

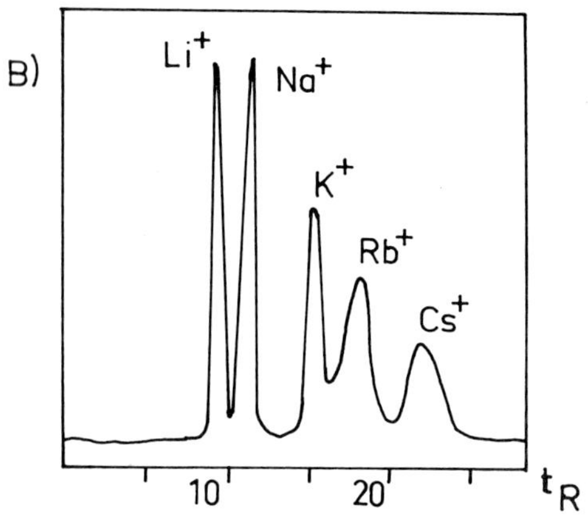

Figure 18 A

Schematic diagram of apparatus for ion chromatography of cations, according to *Small, Stevens*, and *Bauman* [367].
V, reservoir for hydrochloric acid (HCl) eluent; P, pump; I, sample injector; R^-H^+, cation exchanger (actual separation column); R^+OH^-, anion exchanger (stripper column) for removal (neutralization) of eluent; K. DZ, conductometric flow-through cell; K, conductometer; R, output to recorder, integrator.

Figure 18 B

Separation of alkali metal ions by ion chromatography, according to *Small, Stevens*, and *Bauman* [367].
Stationary phase: styrene-divinylbenzene copolymer ion exchanger (H^+ form), column dimensions 250 × 9 mm.
Stripper column: Dowex 1 × 8 (OH^-), column dimensions 270 × 9 mm.
Mobile phase: 0.01 N HCl.
Flow rate: 160 $cm^3\ h^{-1}$ (t_R in min).
Sample: 0.1 cm^3 (0.01 mol/dm^3 of each alkali chloride).
Detection: flow-through conductometer.

background of the mobile phase is eliminated by combination of the separating column with a further column (the stripping column) which neutralizes the eluents. Thus hydrochloric acid frequently used for cation separations can be neutralized by conversion into water with the aid of an ion exchanger in the OH form and the Cl form of the exchanger. The applications of this modern technique have already been

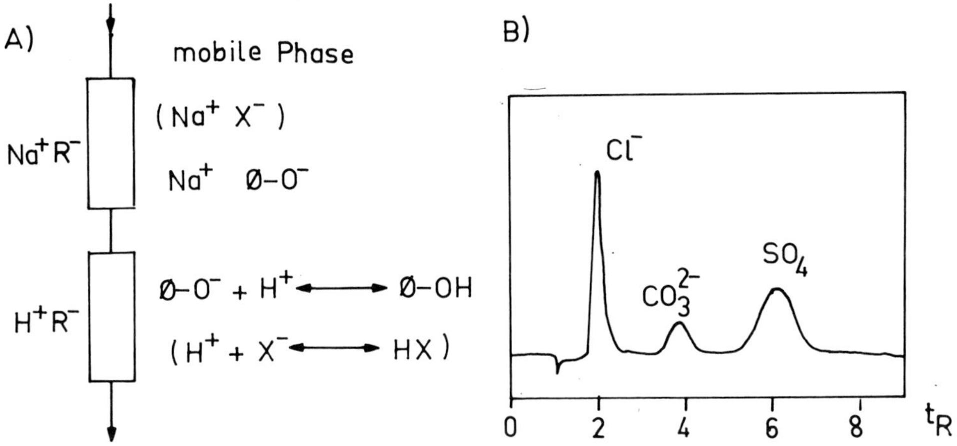

Figure 19 A

Principle of anion chromatography with sodium phenoxide solution as mobile phase, according to *Small, Stevens*, and *Bauman* [367].
Ions and their reactions on use of phenoxide-containing eluents:
Na^+R^-, ion exchanger for separation of anions; $Ph-O^-$, phenoxide ions; H^+R^-, stripper column.

Figure 19 B

Separation of anions in river water by ion chromatography, according to *Small, Stevens,* and *Bauman* [367].
Stationary phase: SA Dowex 2 exchanger (column dimensions 300 × 2.8 mm).
Stripper column: Dowex 50 W × 8 (H^+ form) (column dimensions 250 × 9 mm).
Mobile phase: 0.015 mol/dm^3 sodium phenoxide.
Flow rate: 64 cm^3 h^{-1}.
Sample: 0.01 ml river water.

reported in a monograph by *Sawicki, Mulik,* and *Wittgenstein* (Editor), in 1978 [342], about analysis of environmental materials; examples are found in Chapter III. The first publication in this field was authored by *Small, Stevens*, and *Bauman* in 1975 [367] and deals with the principles of the method (see Fig. 18A and B, and 19A and B).

1.2.5 Gel Chromatography

Table 11

Column chromatography of metal complexes and inorganic ions on gel materials

Compound	Elements investigated	Ref.
Acetylacetonates	Co, Cr	[448, 340]
Acetylacetonates	Co, Fe, Cr, Al, Cu, Ni, Be	[339]
Dithizonates	Various metals	[86]
Acetylacetonates	Cr	[384, 385]
Thiothenoyltrifluoroacetylacetonates	Various metals	[387]
Chloro complexes	Pt, Pd	[256]
Inorganic salts		[40]
Anions		[85]
Phosphates	Mg, Cu	[451]
Pyro-, tri-, and tetraphosphates	Mg	[226, 450]

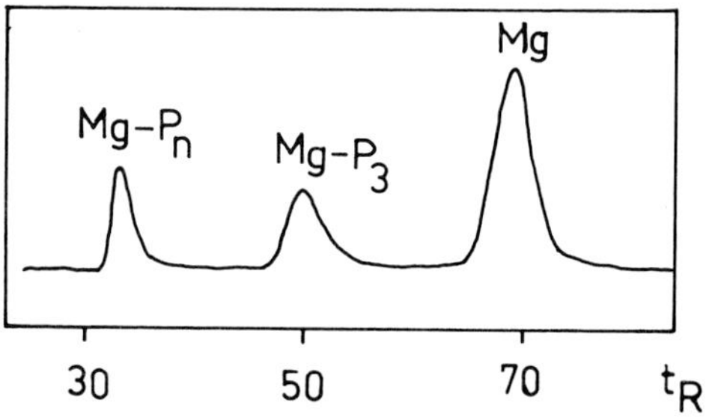

Figure 20

Gel chromatography of magnesium phosphates, according to *Konchiyawa, Yoza*, and *Ohashi* [226].
Mg, Mg^{2+} ions; Mg – P_3 magnesium triphosphate complex; Mg – P_n, magnesium polyphosphate (Mg complex with *Kurrol*'s salt $(KPO_3)_n$).
Stationary phase: Sephadex G 25, column dimensions 975 × 15 mm.
Mobile phase: 0.08 mol/dm^3 NH_3; 0.02 mol/dm^3 NH_4Cl; 1.03 × 10^{-5} mol/dm^3 $MgCl_2$ (pH 10).
Sample: P_3, 0.101 μmol; P_n, 0.495 μmol (based on PO_3 units); Mg 0.515 μmol.
Detection: AAS at 285.2 nm (flame AAS, direct).

During the last five years, attempts have been made to solve specific separation problems of inorganic analysis by modern gel chromatography on various materials. The method is particularly suitable for inorganic salts such as condensed phosphates. The publications listed in Table 11 are still of only theoretical interest for the chromatography of metal chelates, and are concerned with such questions as the dissociation of dithizonates [86] and the partitioning of acetylacetonates [339, 340]. This method has not yet acquired any importance beyond the separation of condensed phosphates.

2 Gas Chromatography (GC)

During the years lying between 1950 and 1960, i.e. the initial years of gas chromatography (GC), not only numerous organic compounds but also inorganic gases and simple volatile inorganic compounds such as halides and hydrides were separated mainly by methods of adsorption GC.

At the beginning of the 1960's, partition GC steadily grew in importance with the increasing number of separation liquids: a large number of new metal chelates were synthesized which could be successfully separated by GC and detected with a variety of sensitive detectors. By 1970/1071, *Rüssel* and *Tölg* [701] could already cite over 600 publications concerned with "Application of gas chromatography to separation and determination of inorganic substances".

The high hopes placed in GC as a method of elemental analysis, and particularly trace analysis, during this early period are now seen to have remained partly unfulfilled. Oligo- or multielement analyses of practical utility have been described in just a few isolated cases. It is not so much a lack of separation capability or performance of this chromatographic method, but rather multifarious problems in the behavior of the comparatively involatile inorganic compounds in the gas phase, and their interactions with the column materials and other components of a gas chromatograph which have imposed these restrictions.

Nevertheless, in conjunction with sensitive and selective detectors, GC has acquired a firm place in the analysis of a series of elements, and especially in trace analysis, as shown by the numerous applications listed in Section III.

The task of GC in organic analysis lies mainly in the resolution of complex substance mixtures into the individual components, the substances themselves usually being volatile already. In contrast, inorganic analysis rather exploits opportunities of removing interfering substances from those to be determined, of increasing selectivity by directed sample preparation (derivatization and extraction), and of optimum preparation of the sample for quantitative analysis in the detector system at the end of the column by gas chromatographic separation.

Table 12 surveys the principal groups of inorganic compounds and substances amenable to GC.

Table 12

Groups of substances for "inorganic gas chromatography".

Group	Examples
A) Gaseous elements, element and metal vapors	N_2, O_2, noble gases Zn, Cd, S, P
B) Inorganic compounds	
Oxides	CO, CO_2, nitrogen oxides
Halides	E.g. of As, Sb, Sn, Ti, U, etc.
Hydrogen compounds	Silanes, boranes, H_2S, NH_3, of As, Sb, etc.
C) Metal chelates	Ligands: β-Diketones, β-thioketones, β-keto amines, salicylaldimines, diethyldithiophosphates, diethyldithiocarbamates Mixed ligands: β-Diketones and dimethylformamide, dibutyl sulfoxide, tributyl phosphate
D) Organometallic and organometalloid compounds	E.g. of elements Si, Ga, Sn, B, P, As, Sb, Bi, Se, Te, Hg, Pb
E) As various organic compounds (after derivatization of inorganic substances)	E.g. triethylfluorosilane (determination of F), iodoacetone (I), 1,2-dibromocyclohexane (Br), piazselenoles (Se), trialkyl esters (or silyl esters) of inorganic acids

Element analysis is also possible if nonvolatile metal chelates undergo reproducible decomposition to volatile substances; e.g. for Ni as polychlorinated xanthate [748], Ru as thiosemicarbazide chelate [470].

Table 13

Books and reviews of gas chromatography in inorganic analysis.

Year	Author	Bibliographic details
1963	*J. Tadmor*	"Gas chromatography in inorganic chemistry", Chromatogr. Rev. **5**, 223. (109 references: separations of gaseous and liquid substances) – Review
1965	*R. J. Juvet* and *F. Zado*	"Inorganic gas chromatography", Advances in chromatography. M. Dekker, New York. – Book
	R. W. Moshier and *R. E. Sievers*	"Gas chromatography of metal chelates", Pergamon, Oxford. – Book
1970	*G. Gnauk* and *H.-J. Berg*	"Gas-Chromatographie in der Metallurgie. Eine Anleitung für die Laborpraxis", Dtsch. Verlag f. Grundstoffindustrie, Leipzig. – Book
1972	*H. Rüssel* and *G. Tölg*	"Anwendung der Gaschromatographie zur Trennung und Bestimmung anorganischer Stoffe", Fortschritte d. chem. Forschung **33**, 1. (629 references: covers the literature up to 1971) – Review
1973	*G. Guiochon* and *G. Pommier*	"Gas chromatography in inorganics and organometallics", Ann. Arbor Sci. Publ., Ann Arbor Mich. – Book
1974	*J. A. Rodriguez-Vazquez*	"Quantitative inorganic analysis by gas chromatography, a review". Anal. Chim. Acta **73**, 1. (190 references: Quantitative analysis, especially applications) – Review
1977	*P. C. Uden* and *D. E. Henderson*	"Determination of metals by gas chromatography of metal complexes, a review". Analyst **102**, 889. (218 references: Separations of metal chelates, detectors, applications) – Review
1979	*G. Schwedt*	Chromatography in inorganic trace analysis. Topics in Current Chemistry **85**, 159. (117 GC-references: trace analysis, applications) – Review

2.1 Adsorption (GC)

Table 14

Separation materials for adsorption gas chromatography of inorganic substances

Separation material	Typical separations	Ref.
Alumina	O_2, N_2, CO_2	[592]
Alumina (+ Na_2MoO_4, NaCl, $LaCl_3$)	Inert gases	[562]
Beryllium oxide	H_2S, H_2O, NH_3	[531]
Silica gel	O_2/N_2, CO_2	[541]
Special silica gel	CO_2, COS, H_2S, CS_2, SO_2	[776]
Silica gel	O_2, O_3	[593]
Silica gel	CO_2, H_2S, SO_2	[639]
Chromium(III) oxide	O_2, N_2, Ar, He	[626]
Clay minerals (Attapulgite, Sepiolite)	O_2, N_2, CO, CO_2	[643]
Clay minerals (Hornite)	O_2, N_2, CO, CO_2	[643]
Kaolin	He, O_2, N_2, CO, CO_2	[530]
Zeolite	H_2, CO	[782]
Sodium-, lithium fluoride, alumina	MoF_6, SbF_5, UF_6, F	[669, 670]
Quartz granules	Ta, Re, Ru, Os, Ir: Oxides, hydroxides	[753]
Sodium tetrachloroferrate(III) ($NaFeCl_4$)	$SbCl_3$	[708]
Chromosorb 102	Element hydrides	[682]
Graphite	NH_3, N_2, H_2	[679]
Synthetic diamond	CF_2O, CO_2	[590]
Carbon molecular sieve	CO_2, H_2S, SO_2	[497]
Molecular sieve	Hydrogen isotopes	[526]
Carbon molecular sieve	O_2, N_2, CO, CO_2, N_2O, SO_2	[538, 756]
Molecular sieve	CO_2, CO, COS, H_2S in helium	[503]
XAD resins	NH_3, SO_2, H_2S, CO, CO_2, H_2O	[553]
Styrene polymer	H_2O, CO_2	[594]
Porapak Q (porous polymer)	H_2S, SO_2, CO_2	[491]
Porapak Q	GeH_4, SnH_4, AsH_3, SbH_3	[609]
Porapak QS polymers	H_2S, CH_3SH, $(CH_3)_2S$, $(CH_3)_2S_x$, SO_2	[534]
Porapak	PH_3	[573]
Porapak Q	Tetramethyltin	[625]
Porapak P	Chlorides of Si, Sn, Ge, P, As, Ti, V, Sb	[661]
Porapak	Alkyl halides	[640]
Porapak Q	ClCN	[654, 655]
Teflon	F, MoF_6, SbF_6, SbF_3	[668]
Sephadex (dextran polymer)	H_2O from organic solvents	[525]

Details of calibration gases: *Brassau*, 1975 [502].

Studies on non-specific effects in adsorption GC for inert gases as examples: *Gerber* and *Sawyer*, 1972 [562].

Adsorption gas chromatography has been used mainly in the separation of gases (see Table 14). The monograph by *Kiselev* and *Yashi*, 1969 [619] considers the fundamentals of this method and its application to the inorganic and organic analysis of gases and liquids. A specialized review on the use of porous polymers is provided by the article written by *Hollis* in 1973 [595] with 39 references.

The principal inorganic and organic separation materials have been compiled in Table 14. Only some of these materials are suitable as stationary phases for quantitative analysis, such as molecular sieves and porous polymers (e.g. Porapak materials, see [491]) which exhibit only weak residual adsorption and only slight tailing of the eluted substances. Interesting possibilities for the separation of hydrogen isotopes are provided by molecular sieves, for example (see Fig. 21). Inorganic substances such as NaF, LiF, and Al_2O_3, as well as the organic polymer Teflon proved suitable for corrosive inorganic fluorine compounds, see refs. [668 – 670].

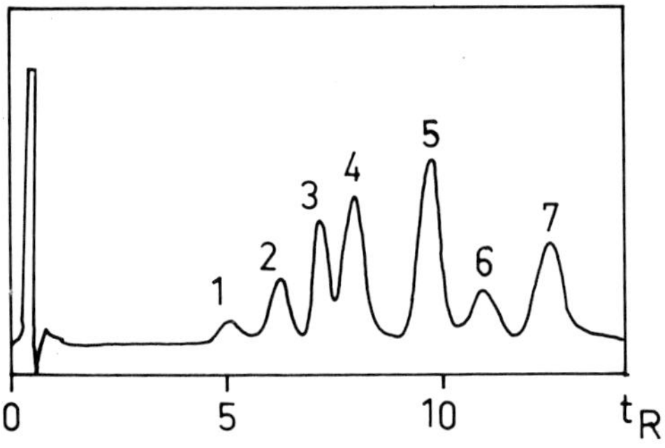

Figure 21

Separation of hydrogen isotopes by gas chromatography on molecular sieves, according to *Conti* and *Lesimple*, 1967 [526].
Separation material: molecular sieve 4A (Union Carbide Corp.), particle diameter 0.25 – 0.31 mm, activated at 250 °C in a helium gas atmosphere (column dimensions 2000 × 2 mm, material copper).
Column temperature: – 148 °C (gas chromatograph fitted with cryostat).
Carrier gas: helium.
Flow rate: 80 cm^3 min^{-1} (t_R in min).
Detector: thermal conductivity cell.
Sample: synthetic mixtures.
1, p-H_2; 2, o-H_2; 3, HD; 4, HT; 5, D_2; 6, DT; 7, T_2.

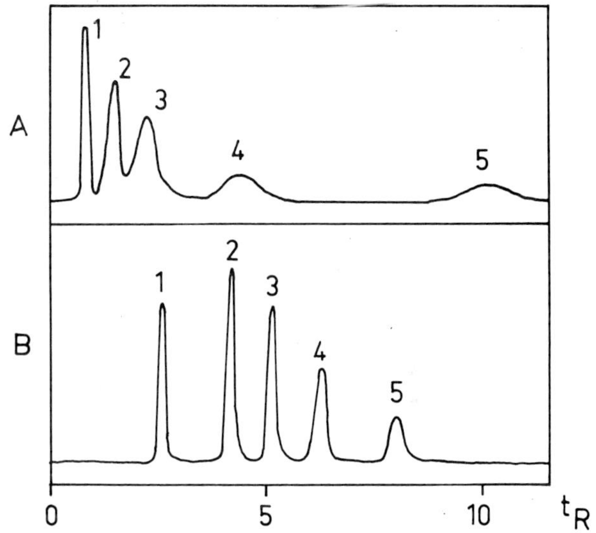

Figure 22

Separation of hydrides of the elements Ge, As, Sn, and Sb by gas chromatography on Porapak, according to *Kardeg* and *Christian*, 1977 [609].
1, air; 2, Ge; 3, As; 4, Sn; 5, Sb.
Separation material: Porapak Q (glass column, 91 cm long).
Column temperature: A, −85 °C; B, temperature program 75 to 120 °C at 8 °C min^{-1}.
Carrier gas: nitrogen (hydrogen).
Flow rate: no details given (t_R in min).
Detectors: mass spectrometer, thermal conductivity cell, atomic absorption spectrophotometer.
Sample: hydrides formed by reduction of hydrochloric acid solutions with $NaBH_4$, volume 5 ml.

The various Porapak materials have been widely applied in the separation of hydrides (see Fig. 22), sulfur-containing compounds, alkyl halides, and also of metal chlorides (see Table 14). Chromium(III) oxide must be activated prior to adsorption chromatography [626], while silica gel must be deactivated after acid treatment outside the column [776]. Various clay minerals (such as kaolinite, hornite, attapulgite, and sepiolite) have been successfully used in the separation of O_2, N_2, CO, and CO_2 (see Fig. 23).

Further applications taken from the literature prior to 1968 can be found in the monograph by *Kiselev* and *Yashi*.

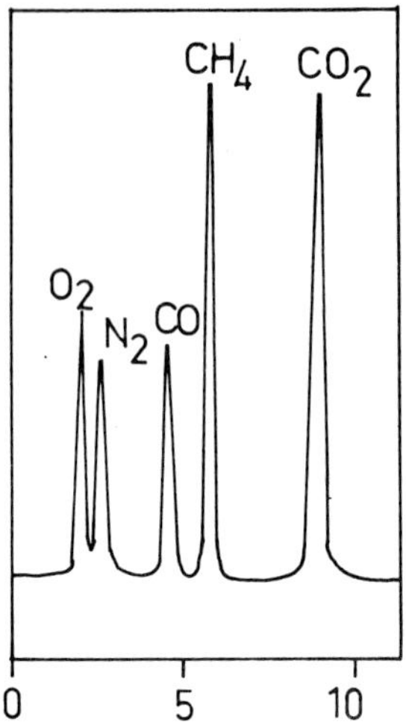

Figures 23

Separation of inorganic gases by gas chromatography on the clay mineral sepiolite, according to *Manara* and *Taramasso*, 1973 [643].
Separation material: Valecas sepiolite, clay mineral of hornite group, 60 – 80 mesh (column dimensions 800 × 4 mm).
Column temperature: – 78 °C to + 70 °C (ballistic program, see [643] for details).
Detector: thermal conductivity cell.
(Lowest plate height: 0.18 cm – separations comparable with those on 5A molecular sieve + 9.5% water.)

Table 15

High temperature gas chromatography (thermochromatography).

Compounds: Elements	Stationary phase/Temperature	Ref.
Bromides: Transactinides	Nickel tube/Gradient 800 – 100 °C	[813]
Chlorides: Te, Nb, Mo, Tc, etc.	Quartz + alkali chlorides/350 °C	[466, 697]
Chlorides (with complex formation): I, Te, Sb, Sn, In, Cd, Zr, Hf, Nb, Ta, Mo, Tc, Re, Ru, Os	KCl, CsCl, NaCl, $BaCl_2$/300 °C	[781]
Oxides, hydroxides: Tc, Re, Ru, Os, Ir, (in O_2- or $O_2 - H_2O$-gas stream	Quartz granules/1500 °C	[753]
Chlorides, oxide chlorides: I, Sb, Mo, Tc, Re, Nb, Zr	Graphite, quartz, quartz + CsCl, $MgCl_2$, KCl/800 °C	[696]
Chlorides, oxides, bromides	Silica gel, graphite/800 °C	[698]
Chlorides	Alkali chlorides/450 – 1000 °C	[777]

In recent years high-boiling halides and oxides and hydroxides have been volatilized at temperatures up to 1500 °C and partly separated in special equipment which is not yet commercially available. This high temperature variant of gas chromatography is also called thermochromatography.

The available evidence is still insufficient to permit any final decision whether this method really does involve chromatographic separation processes (adsorption processes) on the packing materials quartz, graphite, and silica gel, or whether it is a kind of particularly effective thermodistillation in a column.

Owing to the high separation temperatures no irreversible adsorption effects occur on use of, say, oxygen and oxygen/water mixtures as carrier gases for oxides and hydroxides in this kind of gas chromatography. Radioactive components have also been separated [460, 697]. Column packings of alkali and alkaline earth chlorides, and quartz impregnated with these materials, permit volatilization and separation of numerous elements in the form of complex chlorides [781].

The papers published so far have had little impact on practical analytical work.

2.2 Partition Chromatography (GC)

2.2.1 Inorganic Substances

Inorganic gases are separated mainly by adsorption GC (see Table 14). A survey of suitable separation liquids for GLC of sulfur-containing gases is given in the article by *Ronkainen* et al., 1973. Organic polymers containing halogens or pseudohalogens have been used for the separation of gases containing sulfur, chlorine, and fluorine (see Table 16).

Particular attention should be paid to exclusion of moisture on gas chromatography of volatile **metal** and **metalloid chlorides** with modestly polar to nonpolar silicone oils, Kel-F oils or waxes. The carrier gas must be especially dried to avoid hydrolysis of the substances to be separated; moreover, residual activities of the support materials must be eliminated by repeated injection of the sample mixture or also by careful deactivation (silylation). The reactivity of chloride mixtures containing much free chlorine towards the column packing was reported by *Parissakis* et al. in 1971 [661]. Low-melting inorganic compounds such as $FeCl_3 \cdot 6\,H_2O$ (m.p. 37.8 °C), sodium/mercury alloys (m.p. 46.7 °C), and metallic gallium (m.p. 29.8 °C) have been used as stationary phases for metal chlorides [778].

Table 16
Partition gas chromatography of inorganic substances (selection).

Compounds	Stationary phase	Ref.
Inorganic gases		
CO, CO_2, $COCl_2$, HCl, Cl_2	Arochlor 1232 (chlorinated bi- and polyphenyls)	[469]
H_2S, SO_2, CH_3SH	H_3PO_4/cyanosilicone XE-60	[505]
Cl_2, ClF, ClF_3, HCl, Cl_2O, OF_2, ClO_4F, ClO_3F, UF_6	Poly(trifluoromonochloroethylene)-oils – Kel-F	[572]
Halides		
Si tetrahalides	Apolar methylsilicone oils	[647]
$AsCl_3$, $SnCl_4$	Halocarbon oil 6-00	[533]
$SiCl_4$, $TiCl_4$	Kel-F-40-polymer wax	[729, 730]
$SiCl_4$, $GeCl_4$, $SnCl_4$, $TiCl_4$	Silicone DC-550, Apiezon L, polyethylene glycol, $FeCl_3 \times 6H_2O$, Na – Hg alloy, metallic gallium	[778]
Tetrachlorides: Si, Sn, Ge, Ti, V, and PCl_3, $POCl_3$, $AsCl_3$, $SbCl_5$	Silicone oil Kel-F, Apiezon L	[661]
$SiCl_4$, $SnCl_4$, $TiCl_4$	Nitrobenzene, silicone oil FF 7100, Apiezon greases N	[801 a]
Hydrides		
Boron hydrides	Silicone OV-17	[738]
AsH_3, PH_3, GeH_4, SiH_4, H_2S, H_2	Silicone oil	[535]
Elements		
Phosphorus	Silicone OV-1, SE-30	[457]
Sulfur	Silicone oil E 301	[517]

The chlorides of the rare earths have been studied by gas chromatography in glass capillary columns up to 250 °C in the form of complexes with aluminum chloride [814].

In the case of mixed **silicon halides**, and for the chlorides of the Group IV elements, relationships have been found between the retention volume and, respectively, the number of Br and I atoms and the boiling points (*Michael*, 1976 [647]; *Tohyama*, 1974 [778]). The chlorides can also be used for quantitative analysis of the elements, especially in industrial products.

Separations of **hydrides** of elements belonging to Groups III to V can be accomplished not only by adsorption GC but also by means of various silicone oils. Boron hydrides have been separated, for example, by use of temperature programming [738].

Phosphorus and **sulfur**, for example, have been separated in **elemental form** on various silicones as stationary phase. The column temperature was 100 – 120 °C for P (separation of phosphorus hydrides possible) and 245 °C for sulfur.

Reports of the gas chromatography of inorganic compounds date mainly from the years before 1970; further references will be found in the 1972 paper by *Rüssel* and *Tölg*.

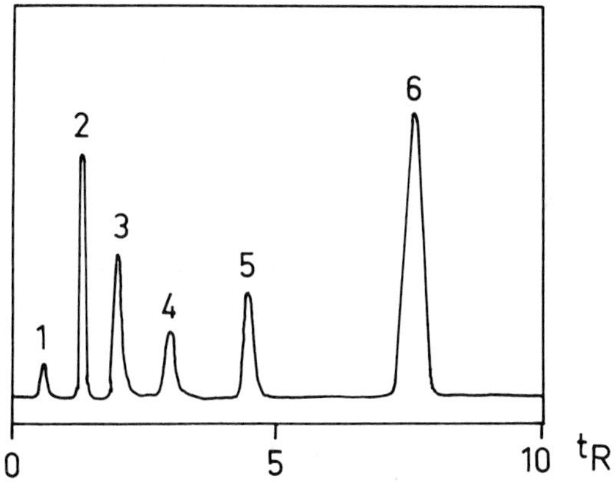

Figure 24

Separation of chlorobromosilanes by gas chromatography, according to *Michael, Danne*, and *Rischer*, 1976 [647].
1, air; 2, $SiCl_4$; 3, $SiCl_3Br$; 4, $SiCl_2Br_2$; 5, $SiClBr_3$; 6, $SiBr_4$.
Separation materials: 5% AK 12500 methylsilicone oil (Wacker Munich) on Fluoropak 80 (PTFE product) (column dimensions, 3000 × 3 mm V2A stainless steel).
Column temperature: 90 °C.
Carrier gas: hydrogen.
Flow rate: 2.5 $dm^3\ h^{-1}$.
Detector: thermal conductivity cell (200 mA).
Detection limit: 1.7 µg $SiCl_4$ – 35 µg $SiBr_4$.

2.2.2 Organometallic Compounds

Table 17

Melting and boiling points of some stable organometallic compounds.

Element	Compound	M.p. °C	B.p. °C
As	Triphenyl-	60	232/14 torr
Bi	Triphenyl-	78	360
Ge	Tetraethyl-	–	163.5
Hg	Dimethyl-	–	92
	Diphenyl-	121	300 (204 at 10.4 torr)
Pb	Tetraethyl-	–	83/15 torr
	Tetraphenyl-	225	–
Se	Diphenyl-	2.5	301
Si	Tetraethyl-	–	153.7
	Tetraphenyl-	234	–
Sn	Tetramethyl-	–	76.6
	Tetraethyl-	–	175
	Tetrabutyl-	–	145/10 torr
	Tetraphenyl-	225	–
Te	Diphenyl-	4.2	174/10 torr

A series of organometallic or organometalloid compounds of main group elements belonging to Groups III to VI, and those of mercury can be volatilized without decomposition (a selection is given in Table 17). In most cases, the alkyl compounds are already gaseous or liquid at room temperature; the aryl compounds are liquids or usually solid.

With regard to the stationary phases employed, gas chromatography of **organometallic compounds** closely resembles that of organic compounds: silicone materials and polymeric esters are particularly suitable (see Table 18).

Work published in this field can be classified into the following three groups:

1) Separations of various organometallic compounds of an element (applications in the preparative organometallic chemistry of germanium, and tin).
2) Quantitative analysis of organometallic compounds occurring in the environment (especially Pb and Hg); see Section III/4.
3) Quantitative elemental trace analysis after conversion of inorganic compounds into organoelement compounds.

Table 18

Gas chromatography of organometallic and organometalloid compounds (selection, see *Rüssel* and *Tölg* for references up to 1971).

Element: Compound	Stationary phase	Ref.
As:		
Alkyl-, arylarsanes, perfluorinated cpds., bromoorganic cpds.	SE-30, Kel-F wax 400	[570]
Triphenylarsane	Silicone DC-200	[599 a]
Triphenylarsane	Carbowax 20 M	[752]
Alkylarsanes	Carbowax 20 M	[763, 764]
Methyl-As cpds.	Carbowax 20 M/2-Nitroterephthalic acid (FFAP)	[761]
Methyl-As diethyldithiocarbamate	OV-17	[532]
Ge:		
Alkylgermanes, -digermanes	Squalane, Carbowax 4000, Silicone 701	[715]
Chalcogen-containing organogermanes	Apiezon L, Carbowax 20 M	[499]
Hg:		
Phenyl-Hg cpds.	Diethyl glycol adipate	[488]
Alkyl-, aryl-Hg	Miscellaneous (survey)	[685]
Diphenyl-Hg	Carbowax 20 M	[752]
Diphenyl-Hg, phenyl-Hg chloride	Ethylene glycol succinate	[638]
Pentafluorophenyl-Hg	Dexsil 300, Duropak Carbowax 400	[651]
Organo-Hg cpds.	FFAP	[761]
Dialkyl-Hg	OV-17, QF-1	[721]
P:		
Aliphatic diphosphites	Distearylpentaerythritol diphosphite, tetrakis(nonylphenyl)polypropylene glycol diphosphite	[811]
Phosphoric esters	OV-1, OV-101	[636]
Pb:		
Alkyl-Pb	OV-101	[678]
Tetraalkyl-Pb	OV-1	[522, 523]
Se:		
Dimethyl selenide, dimethyl diselenide	OV-1	[521]
Alkyl selenides	Polymetaphenyl ether (5-membered ring), Carbowax 20 M, silicone oil DC 550	[543]

Table 18 (Continued)

Element: Compound	Stationary phase	Ref.
Si:		
Methylfluorosilanes, methylpoly-siloxanes	SE-30	[583]
Alkylsilanes	Silicone oils	[671]
Methyl(propyl)dimethylcyclo-siloxanes	Methyltrifluoropropylsiloxane	[807]
Tetraalkoxysilanes	Cyanosilicone GE XE-60	[542]
Tetraalkoxysilanes	Cyanosilicone and Apiezon L	[663]
Tetraalkoxysilanes	Cyanosilicone and Apiezon M	[664]
Silatranes	Apiezon L, OV-17, OV-225	[720]
Sn:		
Butyl-Sn cpds.	Silicones, Reoplex 400 (poly-propylene glycol adipate), Carbowax (polyethylene glycol)	[559]
Butyl-, octyl-, phenyl-Sn halides	Silicone oil	[779]
Organo-Sn cpds.	Apiezon L, neopentyl glycol succinate	[676]
Phenyl-Sn cpds.	OV-17	[735]
Phenyl derivatives		
Se, Te, Hg, As, Sb, Bi	Carbowax 20 M	[752 a]
Methyl derivatives		
As, Sb, Se, Sn	Silicone rubbers, Na dodecyl benzenesulfonate	[641]
Various organometallic compounds		
Ge, Sn, Si	Apiezon L, SE-30, QF-1, XE-1112, o-nitrotoluene	[557]
Metal carbonyls		
Fe, Cr, Mo, W	Squalane/Apiezon L	[672]
Methylbenzenetricarbonyl-Cr complexes	SE-30	[617]

As a consequence of cyclic processes occurring in the environment, inorganic compounds of the elements As, Hg, Pb, and Se are transformed into organic (primarily methylated) derivatives which are far more toxic than the original inorganic substances. In combination with sensitive, selective detectors, gas chromatography represents an extremely powerful method for analysis of these substances.

Table 19

Alkylation and arylation reagents for gas chromatography of organometal and organometalloid compounds.

Reagent	Derivatization	Ref.
Magnesiumdiphenyl	As diethyldithiocarbamate to triphenylarsane	[752]
Tetramethyltin	$Hg^{2+} + (CH_3)_4Sn = CH_3HgCl + (CH_3)_3SnCl$	in [685]
Phenylmagnesium bromide	Se-, Te-, Hg-, As-, Sb-, Bi diethyldithiocarbamates to element phenyls	[752a]
Phenylmagnesium bromide	As-, Sb thionalide to triphenyl cpds.	[764]
Sodium tetraphenylborate	$4\ HgCl_2 + NaBPh_4 + 3\ H_2O = 3\ HCl + NaCl + B(OH)_3 + PhHgCl$	[638, 810]
Phenylsulfinic acid	$PhSO_2H + HgCl_2 = PhHgCl + SO_2 + HCl$	[605]
Lithium pentafluorobenzenesulfinate	$Ph-SO_2Li + Hg^{2+} = (Ph-SO_2Hg^+) = SO_2 + Ph-Hg^+$	[651]
4,4-Dimethyl-4-silapentane-1-sulfonate	$Hg^{2+} + (CH_3)_3-Si-(CH_2)_3-SO_3Na$ to CH_3Hg derivatives	[606]
Pentacyanomethylcobaltate(III)	$Hg^{2+} + (Co(CN)_5-CH_3)^{3-}$ to $HgCH_3^+$	[810]

Inorganic substances, especially halides, can be treated with various alkylation and arylation reagents (Table 19) and then determined by gas chromatography. Use of, say, reagents 1 and 3 (Table 19) requires working in nonaqueous media. This is why the elements As, Se, Te, Hg, Sb, and Bi are first extracted as diethythiocarbamates from aqueous solutions [752, 752a] or, in the case of As and Sb, precipitated as thionalides [764], and then, after removal of the organic solvent or after drying, respectively, reacted with diphenylmagnesium or the Grignard reagent phenylmagnesium bromide (diethyl ether solvent).

Various reagents for methylation or phenylation of aqueous chloride solutions are suitable for derivatization of mercury (see *Zarnegar* and *Mushak*, 1974 [810]). In the case of this element, reaction may occur on gas chromatography between the support and, in particular, diphenylmercury to give mercury(II) chloride and phenylmercury chloride (see *Luckow* and *Rüssel*, 1978 [638]). Decomposition in the column has also been observed with triphenylbismuthane, whereas triphenylarsane is suitable for trace analysis by gas chromatography [752a].

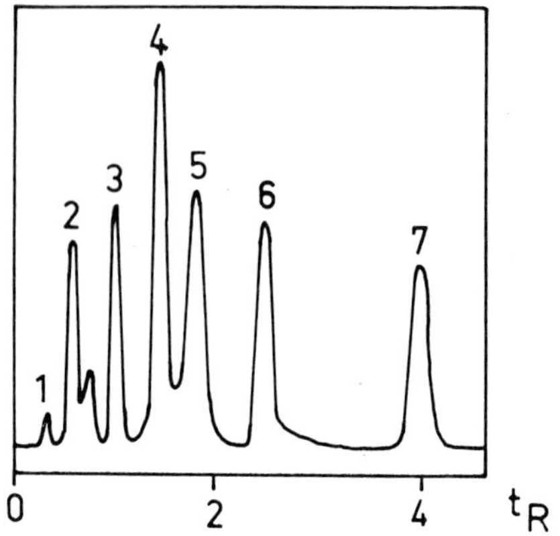

Figure 25

Separation of alkylarsanes by gas chromatography, according to *Talmi* and *Bostick*, 1975 [763].
1, air; 2, AsH_3; 3, CH_3AsH_2; 4, $(CH_3)_2AsH$ + $C_2H_5AsH_2$; 5, $(CH_3)_3As$; 6, $C_3H_7AsH_2$; 7, $C_4H_9AsH_2$.
Stationary phase: 5% Carbowax 20 M on Chromosorb 101, 80/100 mesh (quartz capillary 250 × 0.5 mm).
Column temperature: 175 °C.
Carrier gas: argon.
Flow rate: 100 cm^3 min^{-1} (t_R in min).
Detector: microwave emission spectrometer, plasma at 5 – 20 torr, wavelength 228.8 nm.
Sample: reduction of alkylarsinic acids with $NaBH_4$; enrichment of alkylarsanes in a cold trap.

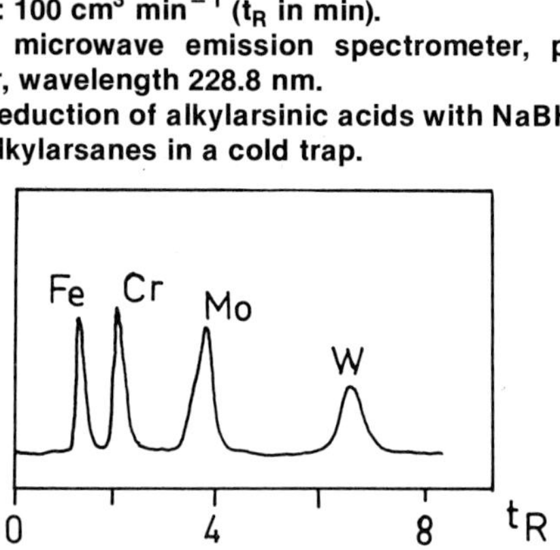

Figure 26

Separation of metal carbonyls by gas chromatography, according to *Pommier* and *Guiochon*, 1970 [672].
Stationary phase: 25% squalane on Chromosorb P (acid washed) 80/100 mesh (column dimensions 1000 × 2 mm).
Column temperature: 90 °C.
Carrier gas: helium.
Flow rate: 33 cm^3 min^{-1}.
Sample: synthetic mixture, 10 µl solution.
$Fe(CO)_5$, $Cr(CO)_6$, $Mo(CO)_6$, $W(CO)_6$.

Organic phosphorus compounds are employed as plant protection agents; phosphoric esters are well suited for gas chromatography (see *Zerba,* 1975 [817]).

Metal carbonyls are likewise volatile; some of them have been separated by gas chromatography (see Fig. 26); isomers can also be resolved.

2.2.3 Metal Chelates

Table 20
Surveys of gas chromatography of metal chelates.

Group of substances	Author/Year	Ref.
Metal chelates	*Kuwamoto*/1972	[633]
Metal chelates	*Stephen*/1972 (13 refs.)	[754]
Metal chelates	*Barnet*/1973 (18 refs.)	[473]
β-Diketonates	*Komanov*/1976	[627]
β-Diketonates	*Burgett*/1976 (162 refs.)	[509]
β-Diketonates (1-5-6*)	*Ross*/1963	[688]
β-Diketonates (3-4-5*)	*Utsonomiya, Shigematsu*/1972	[792]
β-Diketonates (1-3-5-7-9*)	*Belcher* et al./1970	[480]
β-Diketonates (3-7-8-9*)	*Belcher* et al./1972	[483]
Fluorinated 1,3-diketonates etc.	*Joshi, Patak*/1977 (401 refs.)	[607]
Metal chelates of divalent transition metals	*Belcher* et al./1973	[485]
Metal 2-picolyl alkyl ketonates	*Kito* et al./1977	[621]
Quantitative analysis of metal chelates	*Uden, Henderson*/1977	[785]

* No. of ligand in Table 21
(for further literature and monographs, see Table 13)

As long ago as the 1960's, the application of gas chromatography to inorganic analysis advanced on a broad front with the preparation of **volatile metal chelates**, and the first monograph on this topic of chelate gas chromatography was published in 1965 (*Moshier* and *Sievers*, see Table 13). This class of substances opened new vistas in extreme trace analysis of elements, on account of their sensitive detection with flame ionization detectors (FID) and electron capture detectors (ECD).

Table 21

Survey of complexing agents for gas chromatography of metal chelates.

No.	Substituents	Name	Abbreviation
A) *β-Diketones*		(Enol form) 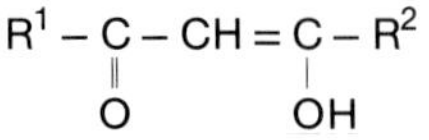	
1)	$R^1 = R^2: CH_3$-	2,4-Pentanedione (acetylacetone)	HAA
2)	$R^1 = R^2: (CH_3)_2CH$-	2,6-Dimethyl-3,5-heptanedione	HDHD
3)	$R^1 = R^2: (CH_3)_3C$-	2,2,6,6-Tetramethyl-3,5-heptane-dione (dipivaloylmethane)	HTHD or (HDPM)
4)	$R^1 = R^2: (CH_3)_2CHCH_2$-	2,2,8-Trimethyl-3,5-nonanedione (isobutyrylpivaloylmethane)	HTND or (HIBPM)
5)	$R^1: CF_3$-, $R^2: CH_3$-	1,1,1-Trifluoro-2,4-pentanedione (trifluoroacetylacetone)	HTFA
6)	$R^1 = R^2: CF_3$-	1,1,1,5,5,5-Hexafluoro-2,4-pentane-dione (hexafluoroacetylacetone)	HHFA
7)	$R^1: CF_3(CF_2)_2$-, $R^2: (CH_3)_3C$-	1,1,1,2,2,3,3-Heptafluoro-7,7-di-methyl-4,6-octanedione (heptafluoropropanoyl-pivaloylmethane)	HFOD or (HHPM)
8)	$R^1: CF_3$-, $R^2: (CH_3)_3C$-	1,1,1-Trifluoro-5,5-dimethyl-2,4-hexanedione (trifluoroacetyl-pivaloylmethane)	HTDH or (HTPM)
9)	$R^1: CF_3 - CF_2$-, $R^2: (CH_3)_3C$-	1,1,1,2,2-Pentafluoro-6,6-dimethyl-3,5-pentanedione (pentafluoro-ethanoylpivaloylmethane)	HPDP or (HPFPD)
10)	R^1: Thenoyl-, $R^2: CF_3$-	1-(Thenoyl-2')-3,3,3-trifluoro-acetone	HTTF
11)		3-Alkylacetylacetone	
12)		β-Diketones with alicyclic substituents	
B) *"Mixed ligands"*			
13)	HTFA (5) + Dimethylformamide		
14)	HTFA + Fluoroorganophosphates		
15)	HTTF (10) + Diethylamine		
16)	HHFA (6) + n-Dibutyl sulfoxide		
17)	HHFA + Tri-n-butyl phosphate		
18)	Decafluoro-3,5-heptanedione + Di-n-butyl sulfoxide		
18a)	Dodecafluoro-4,6-octanedione + Dibutylphosphine oxide, dibutyl phosphate		
C) *Monothio-β-diketones*		(Thiol form) 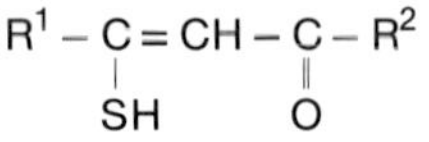	

Table 21 (Continued)

No.	Substituents	Name	Abbreviation
19)	$R^1 = R^2: CH_3$-	Monothioacetylacetone	H(TAA)
20)	$R^1: CH_3$-, $R^2: (CH_3)_3C$-	Monothioacetylpivaloylmethane	H(TAPM)
21)	$R^1: CF_3$-, $R^2: CH_3$-	Monothiotrifluoroacetylacetone	H(TTFA)
22)	$R^1: CF_3$-, $R^2: (CH_3)_3C$-	Monothiotrifluoro-acetylpivaloylmethane	H(TTPM)
23)	$R^1: C_2F_5$-, $R^2: (CH_3)_3C$-	Monothiopentafluoro-propanoylpivaloylmethane	H(TPPM)
24)	$R^1 = R^2: CF_3$-	Hexafluoromonothioacetyl-acetone	H(THFA)
D) *β-Keto amines* (Bidentate ligands)		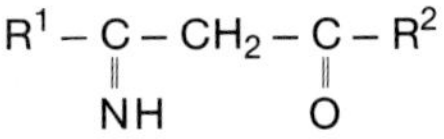 (Imino form)	
25)	$R^1 = R^2: CH_3$-	4-Iminopentan-2-one (aminoacetylacetone)	H(AAA)
26)	$R^1: CF_3$-, $R^2: CH_3$-	4-Imino-5,5,5-trifluoropentan-2-one (aminotrifluoroacetylacetone)	H(ATFA)
27)	$R^1 = R^2: CH_3$- (N-substituted)	4-Methyliminopentan-2-one	H(MAAA)
	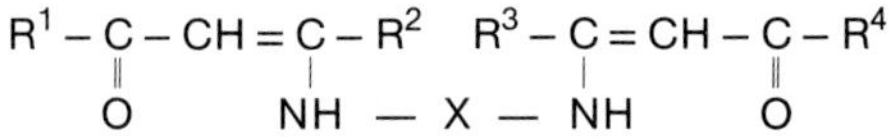		
(Tetradentate ligands)			
28)	$R^1 = R^2 = R^3 = R^4: CH_3$- X: $-CH_2-CH_2-$	Bis(acetylacetone)ethylenediimine (and fluorinated analogues)	
29)	$R^1 = R^4: CH_3$-, $R^2 = R^3: (CH_3)_3C$-, X: $-CH_2-CH_2$-	Bis(acetylpivaloylmethane)-ethylenediimine	
30)	$R^{1,4}: CF_3$-, $R^{2,3}: CH_3$- X: $-CH_2CH_2CH_2-$	N,N′-2,3-Butylenebis(trifluoro-acetylacetonimine)	
31)	R^1 to $R^4: CH_3$-, X: $-CH_2CH_2CH_2CH_2-$	N,N′-2,3-Propylenebis(acetyl-acetonimine)	
E) *Dithiocarbamates and dithiophosphates*			
31)	Dialkyldithiocarbamates	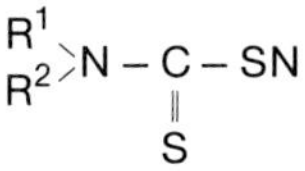	
31 a)	Diethyl-		
31 b)	Trifluoroethyl-		
32)	Dialkyl dithiophosphinates		
F)			
33)	N-Alkylsalicylaldimines		

Table 22
Gas chromatography of metal chelates.

Ligand	Stationary phase/ Temperature (°C)	Elements	Ref.
1)	Apiezon L/170	Be, Al	[490]
	Dow Corning silicone oil 710 R/90 – 215	Cr, Al	[688]
2)	Silicone/200 – 270	Be, Al, Cr, Ni, Fe, Cu, Co, Lu, Er	[722]
3)	Apiezon L, SE-30/250	Mg, Ca, Sr, Ba	[750]
	SE-30, Dow Corning/160 – 280	Nd, Gd, Er	[792]
	Apiezon H/157	Rare earths	[539]
	Silicone XE-60	Cerium group	[620]
4)	SE-30, Dow Corning/160 – 280	Nd, Gd, Er	[792]
	Dow Corning/210 – 280	Sc, Lu, Er, etc.	[791]
5)	DC (Dow Corning) 710/85 – 160	Se, Cr, Cu Mn, Zr, Hg, Zn, Al, Ga, In, Be, Tl	[749]
	Dow Corning/160 – 280	Nd, Gd, Er	[792]
	SE-30 DC-710/110 – 180	Al, Cr	[561]
	SE-30/154	Cr, Al, Rh	[692]
	SE-30/30 – 110	Co, Ru	[794]
	Polyethylene glycol, OV-170/240	Ho, Dy, Tb, Gd	[554]
	Squalane, Apiezon L, DC-710/105 – 120	V	[600]
6)	SE-30/154	Cr, Al, Rh	[692]
	DC 710 R/90 – 215	Cr, Al	[688]
	DC-200/67	Fe, Cr, Rh	[588]
7)	Squalane, Apiezon L, QF-1	Al, Cr, Fe, Cu, Pd	[804]
	Apiezon U	Sc, Yb, Ho, Dy, Tb, Lu	[477]
	SE-30/170	Fe, Ni, Pd, Cr, Cu, Al, Y, Be	[733]
	DC-QF-1/170	Th, U	[547, 548]
	Apiezon L/160	Al, Cr, Fe	[480]
8)	Apiezon L/120 – 170	Be, Al, Fe, In, Cu, Th, rare earths	[767]
	Silicone grease/200 – 280	Rare earths	[723]
	SE-30/200	Rare earths	[765]
9)	Silicone rubber E 301/230	Mg, Ca, Sr, Ba	[483]
	Apiezon L/160	Al, Cr, Fe	[480]
10)	DC 710/270	Be, Al, Sc, V, Cr, Co, Ni, Cu	[602]
11)	Microwax/190	Be, Al, Cr, Cu	[623]
	SE-52, Apiezon/125 – 200	Cu, Cr, Co	[796]
12)	no details	Cu, Cr, Er	[622]
13)	Silicone XE-60/100	Ni, Co	[601]
14)	SE-30/155 – 180	U, Th	[648]
15)	Silicone DC-710/260	Ni, Co	[603]

Table 22 (Continued)

Ligand	Stationary phase/ Temperature (°C)	Elements	Ref.
16)	SE-30, QF-1/245	U, Th	[732]
	OV-1, SE-30, Dexsil 300/170	Mn, Fe, Co, Ni, Zn	[657]
	Dexsil 300	Zn, Cd, Pb	[659]
17)	SE-30/160 – 190	Rare earths	[514]
18)	Dexsil 300/170 – 215	Ce group	[512]
	Dexsil 300/130 – 220	Fe, Co, Ni	[507]
	Dexsil 300/160 – 190	Yttrium group	[510]
18 a)	SE-30, QF-1/185 – 215	Lanthanides	[731]
19)	Apiezon/240	Co, Ni, Pd	[755]
	Apiezon L/170 – 220	Co, Ni, Pd	[482]
20 – 23)	Apiezon L/170 – 200	Zn, Co, Ni, Pd	[479, 484]
24)	SE-30/100, OV-17/80	Cu, Zn, Fe, Ni, Pt, Pd, Cd, Pb	[475]
D)	SE-30/210, 220	V, Cu, Ni (various tetradentate ligands)	[536]
	SE-30/230	vide supra: Cu, Ni, Zn, Cd, Pd, Pt	[662]
25)	Apiezon L, Silicone polymer/ 150 – 220	Cu, Ni	[650]
26)	Silicones/200 and above	Cu, Ni, Pd, Pt	[485]
27)	Apiezon L, Silicone polymers/ 150 – 220	Cu, Ni	[650]
28)	Apiezon L/250	Ni, Pd, Cu, Pt	[486]
	Silicones/200 and above	Ni, Cu	[788, 789]
29)	SE-350/230 – 290	Ni, Cu, Pd	[487]
30)	Dexsil 300/200	Cu, Ni, Pd	[784]
31)	Apiezon L/250 QF-1	Cu, Ni, Pd	[783]
	SE-30/270	Ni	[629, 630]
	SE-30/240 – 270	Zn, Ni	[644]
	OV-25/220 – 265	Zn, Cd, Cu, Ni, Pb	[771]
	GESE-30, Dexsil/240 – 260	Zn, Cu, Ni	[560]
31 a)	OV-101/205 – 245	Zn, Cd, Ni, Pd, Pt	[520]
31 b)	OV-25/120 – 210	Zn, Cu, Ni, Cd, Hg, Co, Fe, Pb, Bi	[770]
	GESE-50/180 – 240	Pb, Bi	[747]
31 a and 31 b)	OV-25/185 – 245	Zn, Cd, Ni, Pb, Sb, Bi	[769]
32)	Apiezon L, SE-30/155	Zn, Ni, Pd, Pt	[519]
	OV-17, Apiezon L, GESE-30, OV-101, OV-25/200 – 220	Zn, Cd, Co, Ni, Pb	[624]
	Silicone GESE/220	Cd	[544]
33)	Dexsil 300/200 and above	Ni, Cu, Zn	[787]

The first chelates were β-diketonates of some metals which could be volatilized even below 200 °C. Incorporation of **fluoroalkyl groups** raises the volatility of the metal chelates, one the one hand, and considerably improves the detection limit for quantitative analysis with ECD, on the other. As a rule, the formation of metal chelates from inorganic substances is simpler than that of organometallic compounds: Aqueous solutions are extracted at a certain pH value with a solution of the chelating agent in an organic solvent. The chelate complex goes into the organic phase, and the excess of reagent can be recovered from the organic phase with dilute sodium hydroxide.

Among the β-diketones, a series of derivatives of acetylacetone (= 2,4-pentanedione) was synthesized and used in gas chromatography of numerous elements (see Tables 21 and 22). The main objective of these studies was to improve the detection limit by raising the number of fluorine atoms (for detection by ECD) and to increase the number of separable elements.

However, gas chromatography of these metal chelates, which are frequently temperature dependent and relatively polar, is not without its difficulties:

In order to minimize decomposition in the column, the support materials must be carefully deactivated (by so-called silanization, i.e. reaction of organic silicon halides with reactive hydroxy groups). Various nonpolar silicones serve as stationary phases at column temperatures up to 280 °C (see Table 22). The use of diketonates for extreme trace analysis nevertheless frequently causes grave problems due to partly irreversible adsorption phenomena and decomposition. New columns can be conditioned, for example, by repeated injection of chelate-containing solutions [786]. Several authors have published papers concerned with losses of metal chelates in the column (see Table 23).

A recent publication by *Sokolow* et al., 1978 [736] gives a detailed account of these **column phenomena**, as investigated for copper(II) trifluoroacetylacetonate with radioactive ^{64}Cu. A silicone stationary phase (SE-54) was used at a column temperature of 120 – 140 °C. The results obtained demonstrate that the sorption of the Cu chelate depends upon the amounts injected, the content of liquid stationary phase in the column, the nature of the support material, and the temperature. They underline the need for both inert support materials and inert stationary phases. Sorption in only partly reversible. *Fontaine* et al., 1975 [549] also reported dissolution and adsorption phenomena in gas chromatography of metal chelates. A reduction of tailing for some metal chelates by addition of the ligand to the carrier gas was reported by *Fujimaya* et al., 1971, 1974, 1976 [554 – 556], and by *Taitiro* et al., 1971 [760].

Table 23

Column phenomena in gas chromatography of metal chelates (examples selected from the literature).

Substance group	Authors
General: Metal chelates	*Rudenko* et al., 1975 [694] (87 refs.)
β-Diketonates	*Uden* and *Jenkins*, 1969 [786], *Burgett* and *Fritz*, 1973 [511]
β-Diketonates: U, Th	*Fontaine* et al., 1972 [548]
Cu trifluoroacetyl-acetonate	*Sokolov*, 1978 [736]
Monothioacetylacetonates	*Belcher* et al., 1971 [482]
Decafluoro-3,5-heptane-dione DBSO chelates of lanthanides	*Burgett* and *Fritz*, 1972, 1973 [510, 511]
Survey of β-diketonates, β-thioketonates, β-keto amides	*Belcher* et al., 1973 [485]

Chelate formation from some elements yields cis-trans isomers, which should not be separated in quantitative analysis of the corresponding element (Cr [632, 760]; Co [722]; Rh [692]).

The aspects of gas chromatography of metal chelates touched upon briefly in this Section are considered in detail and discussed at length in the review by *Uden* and *Henderson*, 1977 [785].

The **volatility of metal chelates** can be determined by thermogravimetry (see Table 24 and Fig. 30) and also by atomic absorption spectrophotometry (*Hildebrand* and *Richett*, 1975 [584]).

Table 24

Thermogravimetry of metal chelates and organometallic compounds – selection of publications.

Substances	Authors
β-Diketonates: Cu, Cr, Co	*Warren* and *Smith*, 1972 [796]
Various β-diketonates: Nd, Gd, Er	*Utsonmija, Shigematsu*, 1972 [792]
β-Diketonates with alicyclic side chains: Cu, Cr, Er	*Kito* et al., 1978 [622]
2-Picolylalkylketonates: Be, Cu, Zn, Ni	*Kito* et al., 1977 [621]
Trifluoroacetylpivaloyl-methanes: rare earths	*Tanaha* et al., 1968 [765]
Fluorinated bis(monothio-β-diketonates): Zn, Co, Ni, Pd	*Belcher* et al., 1972 [484]
Fluorinated alkanoyl-pivaloylmethanes: rare earths	*Belcher* et al., 1972 [483]
Phenyl-Hg(II) compounds	*Belcher* et al., 1976 [488]

Ternary complexes of metal ions, fluorinated β-diketones, and organic solvent molecules, such as dimethylformamide, fluorophosphates, diethylamine, dibutyl sulfoxide, frequently exhibit synergic effects in the separation of the metals from aqueous solution by liquid/liquid extraction. The gas-chromatographic properties are also frequently more favorable than those of the binary complexes (*O'Brien*, 1976 [657], see also *Uden* and *Henderson*, 1977 [785]).

Other suitable ligands have been prepared from **β-thioketones and β-keto amines** and used in gas chromatography. In the case of β-keto amines, tetradentate ligands have been synthesized and used as chelating agents (see Table 21). The same problems attend the gas chromatography of these systems as have already be considered for diketonates. In spite of the wide scope for separation of synthetic mixtures, the analytical applications of the metal chelates have remained limited to individual or just a few elements (see Section III).

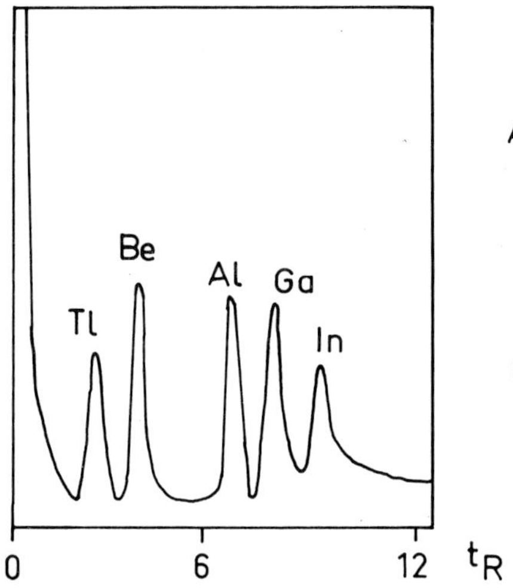

Figure 27

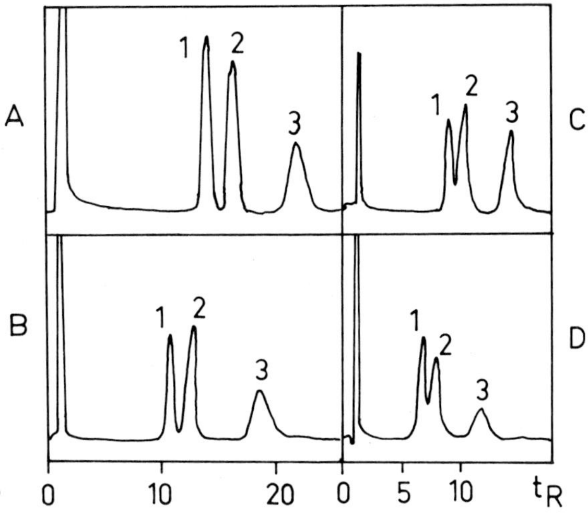

Figure 28

Figure 27

Gas chromatography of metal trifluoroacetylacetonates, according to *Schwarberg* and *Moshier*, 1964 [749].
Stationary phase: 0.5% silicone oil 710 on micro glass beads 60/80 mesh (glass column 1220 × 4 mm).
Column temperature: program from 85 – 160 °C (7.9 °C/min).
Carrier gas: helium.
Flow rate: 79 cm^3 min^{-1} (t_R in min).
Detector: thermal conductivity cell (with tungsten wires).
(Temperature of injection block, 175 °C; detector block, 150 °C.)

Figure 28

Gas chromatography of some metal chelates with tetradentate β-keto amine ligands, according to *Dilli* and *Patsalides*, 1977 [536].
Ligands: A, N,N′-ethylenebis(pivaloylacetone imine); B, N,N′-ethylenebis(isobutyrylacetone imine); C, N,N′-1,2-propylenebis(acetylacetone imine); D, N,N′-ethylenebis(acetylacetone imine).
Stationary phase: 3% SE 30 on Chromosorb 750 80/100 mesh (borosilicate glass column 914 × 8 mm o. d.).
Column temperature: 200 °C.
Carrier gas: nitrogen.
Flow rate: 35 cm^3 min^{-1}.
Sample: 1 μl synthetic mixture.

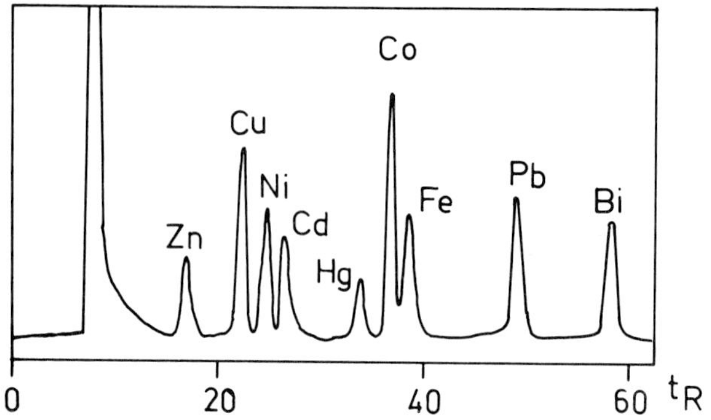

Figure 29

Gas chromatography of metal di(trifluoroethyl)dithiocarbamates, according to *Tavlaridis* and *Neeb*, 1978 [769].
Stationary phase: 3% OV-25 on Chromosorb W-HP 100/120 mesh (column dimensions: 900 × 2 mm).
Column temperature: programmed from 120 – 210 °C (at 2 °C/min). Temperature of injection block/detector block: 210 °C.
Carrier gas: nitrogen.
Flow rate: 35 cm^3 min^{-1} (t_R in min).
Detection: flame ionization detection.
Sample volume: 1 μl.
Sample: extracts of 100 ml of aqueous solution of pH 3 – 4 having following contents [ppm]: Zn, 0.65; Cu, 0.63; Ni, 0.57; Cd, 1.1; Hg, 1.0; Co, 0.58; Fe, 0.55; Pb, 1.0; Bi, 1.1; extracted with 4 × 1 ml of chloroform; chelates dissolved in 1 ml of chloroform.

Considerable recent interest has been focused on **less volatile metal alkyldithiocarbamates** and **-dithiophosphinates** in multielement analysis by gas chromatography. These compounds have been successfully used since the 1950's for the extraction of metal ions and their photometric determination, and for enrichment of trace elements from aqueous solutions for other spectroscopic methods of determination. 1975 saw the first report of gas chromatography of Zn and Ni dithiocarbonates at column temperatures of 240 – 270 °C. The number of separable elements increased in subsequent publications, primarily due to the introduction of fluorinated ligands. However, this method has so far only been applied to the determination of Zn, Cu, and Ni as diethylthiocarbamates in marine sediments (*Radecki* et al., 1978 [677]). The other investigations are limited to extracts of synthetic aqueous solutions. The practical applicability of the technique therefore cannot be predicted, although numerous results of systematic and fundamental studies are already available. Apart from the case of As diethylthiocarbamates (see *Daughtrey* et al., 1975 [532]) nothing has been reported about column phenomena, although such effects as for diketonates are again to be expected.

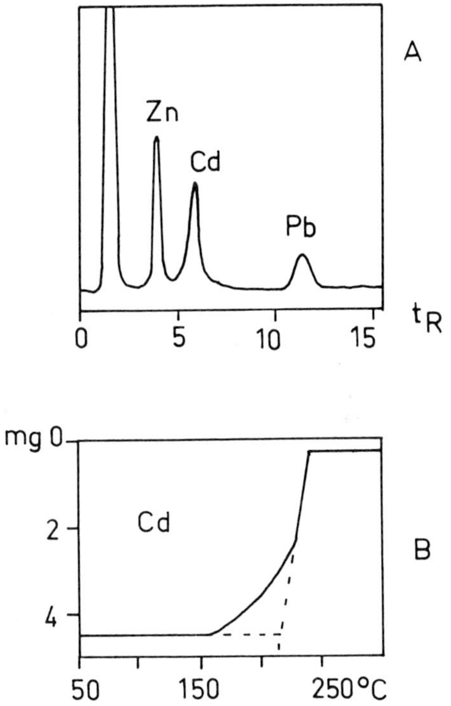

Figure 30

Gas chromatogram of metal diethyldithiophosphinato chelates and thermogravimetric curve of Cd chelate, according to *Kleinmann* and *Neeb*, 1977 [624].
Stationary phase: 3% GE SE 30 on Chromosorb W/AW DMCS 100/120 mesh (column dimensions 780 × 2.9 mm).
Column temperature: 200 °C.
Carrier gas: nitrogen (37 cm^3 min^{-1}).
Detector: FID.

Metal complexes resembling those of keto amines are formed by **salicylaldimine** (see *Uden* and *Henderson*, 1977 [785]). The separation of Cu, Ni, and Zn chelates has been reported; however, their potential in gas chromatography is limited by their lack of thermal stability.

Table 25
High pressure gas chromatography.

Substance group (Elements)	Pressure/Temperature/ Stationary phase	Ref.
Metal acetylacetonates (Ru, Co, Rh, Ir, Pd, Pt)	55 – 69 bar/115 °C/ Epon 1001	[613]
Salicylaldoxime complexes, thenoyltrifluoroacetonates, etc.	63 – 69 bar/130 °C/Kel-F	[614]
Etioporphin-metal chelates (Mg, Cu, Co, Ni, V, Ti, Mn, Zn, Pt, Pd, Ag, Fe) (carrier gas: CCl_2F_2)	30 – 118 bar/145 °C/ Epon 1001, Silicone XE-60	[612]

Reviews:

1) *I. Dovan*, "Supercritical chromatography", Proc. Anal. Chem. **11** (1974) 117 [537].
2) *A. Przjazny* and *R. Staszewski*, "Hochdruck-Gas-Chromatographie und Flüssigkeits-Chromatographie im superkritischen Zustand", Chem. Anal. (Warsaw) **19** (1974) 3; abstr.: Z. Anal. Chem. **272** (1974) 52 [674].
3) *B. A. Rudenko* et al., "Elution chromatography with vapours as mobile phases", Zh. analit. Khim. **30**, 1191 (1975); abstr.: Anal. Abstr. **30** (1976) 6 J 16, (87 references) [694].

At the end of the 1960's a new gas chromatographic method was described for such thermally rather unstable and above all involatile chelates; it uses **dichlorodifluoromethane as mobile phase** at a **pressure** of 30 to about 120 bar and column temperature under 150 °C. Under these conditions the mobile phase may behave as a liquid so that effects known for liquid chromatography may occur. Some metal chelates decomposing at higher temperatures under normal pressure have been separated by this technique of **high pressure gas chromatography** (see Table 25). On account of the carrier gas used, detection is by AAS. The method has not acquired analytical importance, since the development of HPLC at the beginning of the 1970's held more promise for these substances.

2.2.4 Derivatives of Anions

Inorganic anions can be separated and determined by gas chromatography only after prior conversion into organic derivatives. In some cases, analysis in the form of inorganic substances is possible, as with iodine (from iodide) or cyanogen chloride (from cyanide), see Table 26.

A number of reactions have been tried for **halides**, of which the formation of phenylmercury chloride (from phenylmercury nitrate at pH 1.5) and the reaction of aqueous fluoride solutions with alkylchlorosilanes to give alkylfluorosilanes have proved of value in trace analysis too. The possibilities of trace-analytical determination of HCl by gas chromatography have been summarized and discussed at length by *Vierkorn-Rudolph* and *Bächmann*, 1979 [795] and *Bächmann* et al., 1978 [468].

Nitrate determinations are made possible by reactions, for example, with aromatic compounds such as benzene and anisole (see *Tesch*, 1976 [775] and *Tanner*, 1979 [768]). Incomplete reaction below 10^{-4} mol/dm^3 can give rise to problems in such determinations.

Anions of oxy-acids can be converted into trimethylsilyl esters and separated by gas chromatography. *Butts* [513] has treated the ammonium salts of various anions in dimethyl formamide with bis(trimethylsilyl)-trifluoroacetamide (see Table 27 for other silylation reagents). Once again, quantitative reactions are often difficult to accomplish. Arsenic and phosphate determinations have been performed in trace analysis [582, 631, 645]. Silicate ions containing up to four silicon atoms can be treated with hexamethyldisilazane in hydrochloric acid solution [806]. Gas chromatography of trimethylsilyl (TMS) esters is usually performed with temperature programming.

Reactions of **selenous acid** with 1,2-diaminobenzenes to give **piazselenoles** have long been used for extractive isolation and spectrophotometric and fluorometric determination of selenium. The applicability of this reaction to gas chromatography was first demonstrated in 1968 with the reagent 1,2-diamino-4-chlorobenzene; an election capture detector (ECD) was used for the corresponding piazselenole derivative (*Nakashima* and *Toei* [653]). Other derivatives of piazselenole have been examined with regard to their gas chromatographic properties and sensitivity of detection (see *Shimoshi*, 1977 [726]). This method has proved to be a very powerful, highly selective technique in a number of studies on the determination of selenium, mainly in environmental samples (see Section III/2.4). Even so, losses occur in the column when working with these derivatives in the picogram range (see *Kaiser* et al. [610]); they can be overcome, however, by column conditioning (repeated injection of samples containing piazselenoles).

Table 26
Gas chromatography of anions.

Anions	Derivatives	Stationary phase (Temperature °C)	Ref.
Fluoride	Triethylfluorosilane	Silicone oil (60/75)	[495, 700]
Chloride	Phenyl-Hg chloride	Diethyl glycol succinate (180)	[481]
Chloride	1,3-Dichloropropanol, 1-chloro-3-bromopropanol, 1-chloro-3-iodopropanol	No details (review)	[795, 468]
Bromide	1,2-Dibromocyclohexane	OV-17 (70 – 170)	[462]
Iodide	Iodine	Dinonyl phthalate (165)	[665, 569]
Iodide	Iodoacetone, iodoketones	SE-30 (120)	[580]
Chloride, bromide, iodide	Halo alcohols	Ethyl glycol succinate (100)	[699]
Cyanide	CNCl	Halocomid M-18 (55)	[793]
Nitrite nitrate	Nitrotrimethoxybenzene, 4-nitrotetrafluoroanisole	SE-30 (100 – 150) Poly-n-phenyl ether (150 – 175)	[775] [768]
O-Containing anions as trimethylsilyl derivatives			
Silicates	(Si_1 – Si_4)	SE-30 (290)	[806]
Phosphate		Methylsilicone oil DC 430 (80 – 220), 2,2-dimethylpropane-1,3-diol succinate polyester (100)	[576 – 578]
Phosphate	Trimethylsilyl esters	OV-225 (65 – 150)	[645]
Phosphate	Trimethylsilyl esters	SE-30 (90 – 170)	[799]

Table 26 (Continued)

Substituents		Stationary phase (Temperature °C)	Ref.
Selenite	Trimethylsilyl esters	Silicone UCW-982 (70 – 150)	[508]
Arsenate	Trimethylsilyl esters	OV-17 (70 – 150)	[631]
Borate, phosphite, phosphate, carbonate, sulfate, arsenate, arsenite, vanadate	Trimethylsilyl esters	OV-17, SE-30 (70 – 150)	[513, 515]
Arsenate, arsenite, diethylarsinic acid	Trimethylsilyl esters	SE-30, OV-Silicones (60 – 150)	[582]
Piazselenoles for gas chromatography of selenium			
4,5: Phenyl		SE-30 (165)	[808]
5: Cl		SE-30 (200)	[653]
5: NO_2		SE-30 (200)	[724]
5: CH_3-, C_2H_5- or C_2H_5O-		SE-30 (120 – 210)	[726]
5,6: Cl-4,6: Br-4: Br, 6: NO_2		SE-30 (120 – 210)	[726]

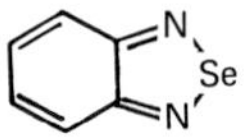

Table 27

Silylating reagents for gas chromatography of anions of oxy-acids as trimethylsilyl esters.

Reagent	Anions (Reaction)	Ref.
Bis(trimethylsilyl)trifluoroacetamide (BSTFA)	NH_4 salts of various anions in dimethylformamide	[513, 515]
Hexamethyldisilazane (HMDS)	Silicate ions in HCl solution	[806]
Methyl(trimethylsilyl)-heptafluorobutyramide (MSHFBA)	$(NH_4)_3AsO_4$	[631]
HMDS + Trimethylchlorosilane (TMCS)	$(NH_4)_2HPO_4$	[577, 799]
BSTFA + TMCS	$(NH_4)_2SeO_3$	[508]
BSTFA	Arsenite, arsenate, diethylarsonic acid	[582]

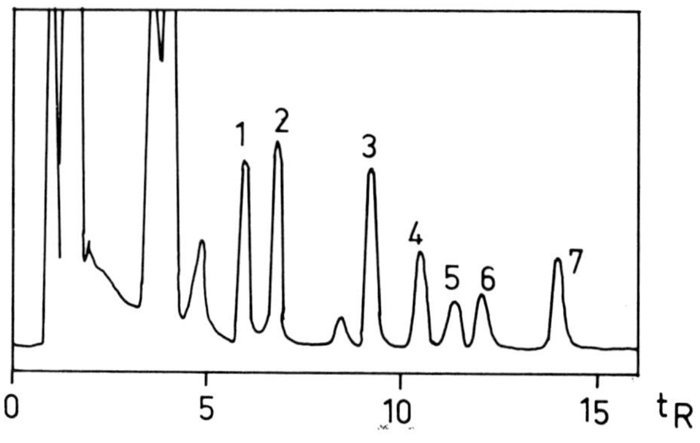

Figure 31

Separation of TMS esters of selected anions by gas chromatography, according to *Butts* and *Rainez*, 1971 [515].
Stationary phase: 3% Silicone OV-17 on Chromosorb W (HP) 80/100 mesh (glass column 3650 × 6 mm).
Column temperature: 70 – 105 °C (at 5 °C min^{-1}).
Carrier gas: helium.
Flow rate: 80 cm^3 min^{-1} (t_R in min).
Detector: flame ionization detector (FID).
1, phosphite; 2, carbonate; 3, oxalate; 4, sulfate; 5, vanadate; 6, phosphate; 7, arsenate.

II Detection Methods

1 Layer Chromatography

1.1 Visual Methods

1.1.1 Colored Substances

Visual detection or instrumental quantitative analysis of colored substances on thin-layer plates constitutes one of the possibilities of detection in layer chromatography. Particularly suitable for such detec-

Table 28

Visual detection of colored substances in layer chromatography (selection).

Ligand, substance group	Elements	Ref.
Dithizonates	Ag, Cd, Co, Cu, Hg, Ni, Pb, Zn	[20]
Dithizonates	Hg, Bi, Cu, Zn, Sn	[446]
Dithizonates	Hg, Cu, Pd, Bi, Ni, Co, Zn, Pb	[386]
Dithizonates	Heavy metals	[336]
Dithizonates	Cu, Co, Ni	[230]
Dithizonates	Cd, Co, Ni, Zn, Cd	[90]
Picrates	Cs, Rb	[165]
2-Hydrazinothiazole	Cu, Co, Ni, Mn	[28, 233]
8-Mercaptoquinoline	Metals of groups I, II, VIII, and IV	[222]
1-Hydroxy-2-pyridine-thione	Metals of group VIII	[223]
4-Pyridin-3-aldehyde 2-quinolylhydrazone	Fe, Ni, Co, Cu	[112]
8-Hydroxyquinoline, 2-methyl-8-hydroxy-quinoline	Al, Be, Cr	[445]
1-Nitroso-2-naphthol	Co, Ni, Mn, Fe	[349]
Diphenylthiovioluric acid	Ni, Co, Cu, Fe, Pd, Pt, Ru	[190]
Dithizonates, oxinates, 2-methyloxinates	Fe, Cu, Hg, Cd, Co, Ni, Mn, Pb, Zn, Bi, Al, Sn, Be, Cr	[447]

tion are metal chelates having absorption maxima in the visible region, which are also used in spectrophotometric analysis. A selection of such compounds for thin-layer chromatography is presented in Table 28.

1.1.2 Reactions on the Layer

Separation of colored ions or of metal complexes absorbing in the UV region can be followed by the application of many spray or locating (metal) reagents, which have already been used for detection of ions in aqueous solutions or after extraction of water-insoluble complexes into organic solvents. With the aid of these reagents, the selectivity of separations can be considerably enhanced by the additional selectivity of detection.

The literature available on this field of detection on layer materials is therefore far more extensive than that concerning the visual detection of colored substances. The following Tables provide systematic

Table 29

Reviews of the use of locating reagents in paper and thin-layer chromatography of inorganic substances.

General publications:

in: "Anfärbereagentien für Dünnschicht- und Papier-Chromatographie", Merck-Darmstadt.

in: *J. C. Touchstone* and *M. F. Dobbins*, "Practice of thin-layer chromatography", Wiley, New York 1978.

For anions:

M. H. Hashmi and *N. A. Chugtai*, 1968 [151]

For cations:

M. H. Hashmi and *R. Chugtai*, 1971 [152]
G. A. Mora, 1966 [292]
K. N. Johri and *H. C. Mehra* – 1970, 1971 [186, 188]
S. Takitani et al., 1966 [405]
U. A. T. Brinkman et al., 1973 [50]
G. Nickless, 1968 [303]

Paper chromatography:

R. A. Webb et al. – 1969 [436]
H. Sulser – 1966 [383]
H. J. Senf – 1966 [347]
J. F. Connolly and *M. F. Maguire*, 1963 [75]

surveys of the pertinent literature (Table 29), the locating reagents used (Tables 30 and 31), and the detection limits attainable for cations (Tables 32 and 33) and for anions (Table 34).

Table 30

Locating reagents for inorganic ions on layer materials after separation by chromatography, according to *Touchstone* and *Dobbins*, 1978 [416].

Reagent	Elements detectable (color)
Alizarin	Numerous ions, including Ba, Ca, Mg, Al, Ti, Fe, Zn, Li, Th, Zr, Se (violet to red spots)
Ammonium molybdate/ tin(II) chloride	Phosphate, phosphite (blue spots)
Ammonium sulfide	Ag, Hg, Co, Ni (black) – Au, Pd, Pt, Pb, Bi, Cu, V, Ti (brown) – Cd, As, Sb (yellow) – Sb (orange)
Aurintricarboxylic acid	Al, Cr, Li
Benzidine	Au(III), Ti(III), CrO_4^{2-}, Mn(IV)
Bromocresol violet	Halides except fluorides
Brucine/2 N NaOH	Bromate (deep red), nitrate (red/yellow), chlorate (red-brown/red)
Quinalizarin/KI	Bi (orange); Ag, Hg, Pb, Sb, V, Te (yellow); Cu (brown); Pt (pink)
Copper sulfate/ mercury chloride/ ammonium thiocyanate	Zn (red/violet), Cu (yellow), Fe(III) (red), Au (orange), Co (blue)
Diacetyldioxime	Ni (red)
Diphenylcarbazide	Ni (blue), Co (orange-brown), Ag, Pb, Cu, Sn, Mn (brown), Zn (violet)
Potassium thiocyanate	Au (orange), Pt, Mo (orange-red), Hg(I) (black), Bi, U (yellow), V, Co (blue), Fe (deep red), Cr (violet), Cu (green-black)
Pyrocatechol violet	Organic tin compounds (dark blue)
Violuric acid	Li (red-violet), Na (violet-red), K (violet), Be (yellow-green), Mg (yellow-orange), Ca (orange), Sr (red-violet), Ba (bright-red), Co (green-yellow), Cu (yellow-brown)

Further spray reagents for paper and thin-layer chromatography:

Reagent	Elements detectable	Ref.
2,2′-Diquinoxalyl	Sn	[15]
$K_3Fe(CN)_6$	U	[1, 26]
Na rhodizonate	Sr	[34]
Bismuthiol II	Te	[314]
Rhodamine G	In	[182]
Salicylaldehyde	Al, Be (fluorometric)	[23]
Aluminon	Cu, Al	[379]
Thiourea	Ru, Os – Se	[334 – 296]
2-Mercapto-5-anilino-1,3,4-thiadiazole	Cd, Bi, Pb	[323]
Rubeanic acid	Ni, Co, Cu	[164]
Pyridin-2-aldehyde 2-quinolylhydrazone	Ni, Co, Cu	[111]
Neocuprin, diacetyl-dioxime, 3,3′-dimethyl-naphthidine	Ni, Co, Cu	[116]
Quercitin	Several elements	[189]
4-(2-Thiazolyl-azo)resorcinol TAR	Co, Cu, Ni	[107]
2-Hydrazinothiazole	Cu, Co, Fe	[233]
Thioacetamide	Cu, Pb, Hg	[273]
1,3,4-Thiodiazole-2-thiol-5-amine	Pt metals	[432]
Arsenazo III	Rare earths	[435]
Neothorin	Rare earths	[201]
Hematoxylin	Organic Sn cpds.	[429]
Bromocresol violet	Halides	[413]
Morpholine-4-carbo-dithiolate	Fe, Co, Ni, Cu, Ru, Rh, Pd, Ir, Pt, Te	[234]
PAR, PAN	Co, Ni, Cu, Bi, Pb, Mn, Cd, V	[290]

The cited publication by *Hashmi* et al. [153] also gives separation procedures following group separations listed in Table 33. Moreover, Rf values and the compositions of locating reagents and TLC separation systems (silica gel and alumina) are to be found therein.

Table 31

Common locating reagents for individual elements, compiled in accord with "Anfärbereagentien für Dünnschicht- und Papier-Chromatographie", Merck, Darmstadt, FRG.

Element	Reagent
Al	Alizarin/aurintricarboxylic acid/ 8-hydroxyquinoline-kojic acid/morin
Sb	Sodium dithionite/sodium thiosulfate/ copper acetate/quercetin
Ba	Na rhodizonate/8-hydroxyquinoline
Pb	Diphenylcarbazide/dithizone/dimercapto-thiodiazole
Cd	Diphenylcarbazone/Titan yellow
Ca	Alizarin/dimercaptothiadiazole/ diphenylcarbazide/8-hydroxyquinoline
Cr	Aurintricarboxylic acid/benzidine peroxide/ quercetin
Fe(III)	Potassium hexacyanoferrate(II)/2-nitroso-1-naphthol-4-sulfone
Ge	Phenylfluorone
Au	Tin(II) chloride-potassium iodide
K	Cobalt lead nitrite/quercetin
Co	Rubeanic acid
Cu	Diphenylcarbazide/sodium sulfide/ quercetin/rubeanic acid
Li	Alizarin/aurintricarboxylic acid/quercetin
Mg	Alizarin/8-hydroxyquinoline/kojic acid
Mn	Benzidine peroxide/diphenylcarbazide/ quercetin/rubeanic acid
Na	Magnesium uranylacetate/zinc uranylacetate
Ni	Quercetin/rubeanic acid
Hg	Sodium sulfide
Se	Alizarin
Ag	Diphenylcarbazide/diphenylcarbazone/ sodium sulfide
Sr	Sodium rhodizonate/8-hydroxyquinoline
U	Pyridylazonaphthol PAN
Bi	Dimercaptothiadiazole/sodium sulfide
Zn, Sn	Diphenylcarbazide
Th, Zr	Alizarin

Table 32

Visually attainable detection limits for various locating reagents and elements, according to *Hashmi* and *Chugtai*, 1971 [152].

Reagent	Element (color)	Detection limit μg
1-Nitroso-2-naphthol	Co (red)	0.01
1,1'-Dipyridyl	Mo(VI) (violet)	0.3
Benzoinoxime	Cu (green)	0.1
Dimethyldiacetyl-dioxime	Ni (red)	0.03
Dithizone	Zn (red)	0.02
Rhodamine	Ag (red-violet)	0.1
Benzidine	Au(III) (gray-black)	0.1
Chromotropic acid	Ti(IV) (red-pink)	0.1
Phenylfluorone	Ge(IV) (orange)	0.05
Tannic acid	W(VI) (brown-yellow)	0.5
Diphenylcarbazide	Sn(IV) (violet)	1.0
8-Hydroxyquinoline	Ga(III) (yellow)	1.0
Rubeanic acid	Ru(III) (blue) – Cu (blue) – Co (blue)	0.05 – 0.02 – 0.02
Na rhodizonate	Ag (brown)	0.5
$K_3Fe(CN)_6$	Fe(III) (blue)	0.04
KSCN	Mo(VI) (red)	0.25
Phosphomolybdic acid	Sb(V) (blue)	0.5
Na_2S	Pb (brown)	0.05
KI	Tl(I) (yellow)	0.2
$(NH_4)_2S_x$	Tl(I) (orange-yellow)	0.5
K_2CrO_4	Tl(I) (orange)	1.0
$SnCl_2$	Au(III) (violet-brown)	0.05
$SnCl_2$/KI	Pt(IV) (yellow-brown)	0.1

Table 33

Locating reagents and visually attainable detection limits for cations, according to *Hashmi* et al., 1966 [153].

Element (cation)	Reagent (color)	Detection limit in μg
Fe(III)	Tannic acid (dark violet)	0.06
Au(III)	Tannic acid (gray-black)	0.3
Mo(VI)	Tannic acid (deep yellow)	0.6
V(V)	Tannic acid (dark blue)	0.9

Table 33 (Continued)

Element (cation)	Reagent (color)	Detection limit in µg
Ga(III)	Dithizone (red-orange)	2.0
Sb(V)	Dithizone (red-orange)	2.5
As(III)	Dithizone (yellow)	6.0
Te(IV)	$SnCl_2$/KI (dark brown)	0.3
Se(IV)	$SnCl_2$/KI (red)	2.0
Rh(II)	$SnCl_2$/KI (yellow)	0.6
Ge(IV)	Phenylfluorone (red)	0.7
Pd(II)	Rubeanic acid (yellow-brown)	0.9
Pt(IV)	Rubeanic acid (red-violet)	0.3
Co(II)	Rubeanic acid (dark brown)	0.2
Cu(II)	Rubeanic acid (red-brown)	0.15
Ru(III)	Rubeanic acid (blue)	0.1
Co(III)	Diphenylcarbazide (red-violet)	0.4
Zn	Diphenylcarbazide (red-violet)	0.3
Sn(IV)	Diphenylcarbazide (red-violet)	0.3
U(VI)	Alizarin (violet)	1.0
Zr(IV)	Alizarin (violet)	1.3
Cr(III)	Benzidine peroxide (dark blue)	1.0
Mn(II)	Benzidine peroxide (blue)	2.8
La(III)	Quinalizarin (violet)	2.0
Al	Quinalizarin (violet)	1.0
Hg(II)	Na_2S (black)	3.0
Cd	Na_2S (yellow)	2.5
Bi(III)	Na_2S (dark brown)	2.0
Pb	Na_2S (brown)	traces
Ni	Diacetyldioxime (red)	0.3
Hn	Diacetyldioxime (brown-yellow)	4.2
Sr	Na rhodizonate (red)	2.2
Ba	Na rhodizonate (red)	2.8
K	Na cobaltinitrite (yellow)	2.0
Rb	Na cobaltinitrite (black-brown)	2.2
Cs	Na cobaltinitrite (orange-brown)	3.5
Li	$AgNO_3$/fluorescein (brown)	0.4
Hg	Alizarin (red-violet)	0.8
Ca	Alizarin (blue-violet)	1.0
Th(IV)	Alizarin (blue-violet)	1.3
Ce(III)	Alizarin (blue-violet)	1.5

Table 34

Locating reagents and visually attainable detection limits for anions, according to *Hashmi* et al., 1968 [153] and *Hashmi* and *Chugtai*, 1968 [151].

Anion	Reagent (color)	Detection limit in µg
AsO_2^-	Silver nitrate (yellow)	0.2
AsO_4^{3-}	Silver nitrate (orange)	0.5
$H_2PO_4^-$	Silver nitrate (yellow)	0.3
CrO_4^{2-}	Silver nitrate (yellow-orange)	0.1
$S_2O_3^{2-}$	Silver nitrate (black)	0.5
Cl^-	Silver nitrate (blue)	1.0
Br^-	Silver nitrate (brown)	0.7
I^-	Silver nitrate (brown)	0.5
S^{2-}	Silver nitrate (black)	0.3
$Fe(CN)_6^{3-}$	Iron(III) chloride (green)	0.3
$Fe(CN)_6^{4-}$	Iron(III) chloride (blue)	0.2
CH_3COO^-	Iron(III) chloride (brown)	2.0
SCN^-	Iron(III) chloride (brown)	0.2
IO_4^-	Potassium iodide/HCl (brown)	1.0
IO_3^-	Potassium iodide/HCl (yellow)	0.2
BrO_3^-	Potassium iodide/HCl (yellow)	3.0
NO_2^-	Potassium iodide/HCl (yellow-brown)	0.2
ClO_3^-	Potassium iodide/HCl (yellow)	2.7
CrO_4^{2-}	Diphenylcarbazide (violet)	0.01
Br^-	Fluorescein (pink)	0.08
I^-	Palladium chloride (brown-black)	0.2

1.2 Quantitative Methods

Survey:

1) **Visual** (semiquantitative) **methods**
   a) for colored substances (absorption in the visible region)
   b) after spraying with locating reagent
   c) fluorescence, diminution of fluorescence (sometimes called fluorescence quenching, under UV light in layers containing a fluorescence indicator)

2) **Planimetry** (measurement of spot area)
   a) for colored substances
   b) after spraying with locating reagent
3) **In situ spectrophotometry (transmission mode,** absorption of transmitted light), photodensitometry
   a) for colored spots or zones
   b) after spraying with locating reagent
4) **Reflectance measurements** (absorption measurements by reflected light)
   a) for colored substances
   b) after spraying with locating reagent
5) In situ **fluorometry**
6) In situ **electrometric measurements**
7) In situ **radiometry**
8) In situ **X-ray fluorescence**
9) **Determinations after elution**
   a) Spectrophotometry
   b) Fluorometry
   c) Emission spectrometry (flame photometry)
   d) Infrared spectrometry (IR)
   e) Activation analysis (neutron)
   f) Titrimetry
   g) Ring oven colorimetry
10) **Determination after digestion**
   (of layer material)

Depending upon the analytical and separation problem at hand, paper and layer chromatograms can be evaluated directly on the layer, i.e. in situ (see items 1 – 8 of survey), after elution (9) of the separated substances, or, after work-up (10) of the cellulose material or the separated complexes, by a very wide range of determination methods from inorganic analysis.

Most commonly used are measurements of transmission (measurement of absorption in transmitted light) and reflectance (absorption measurements in reflected light) of colored substances. Both methods permit reproducible evaluations with standard deviations down to below 1%. The advantages, and disadvantages of these two methods for direct quantitative photometry on thin-layer chromatograms are considered in detail in the publication by *Hezel*, 1973 [165a].

Several TLC evaluation methods, such as a) determination of spot size, b) measurement of absorption in transmitted light, and c) radiometric measurements on radioactive isotopes for the separated cations of the elements Na, K, Mg, Cu, Co, Ni have been compared with regard to practical execution and to results by *Seiler*, 1967 [543] (see also Table 36).

Table 35

Methods of quantitative layer chromatography of inorganic substances (examples)*.

A, adsorption; C, cellulose; S, silica gel; P, partition

1) Visual (semiquantitative) methods
a) for colored substances

Element: Compound	Chromatography	Detection limit	Ref.
Fe, Co, Mn, Ni, Zn, Co, Rh, Pb, Cd, Hg: Tetraphenylporphyrin chelates	TLC/A	10^{-10} mole	[176]
Cu, Ni, Cr, Co, Fe, U, V, Mo: Diethyldithiocarbamates	TLC/A	2 – 300 ng	[216, 398]
Cu, Co, Ni, Pd, Rh: Monothio-β-diketonates	TLC/A	Co: 10 ng, otherwise 50 ng	[294]
Co, Cu, Ni, Fe: Pyridin-3-aldehyde 2-quinolinylhydrazone chelates	TLC/A	7 – 10 ng	[112]
Co, Ni, Mn, Fe: 1-Nitroso-2-naphtholates	TLC/A	60 – 100 ng	[349]
Co, Ni, Fe, Mn, Cu, Bi: Salicylaldoxime compounds	TLC/A	1 – 1.5 μg	[351]
Cd, Pb, Cu, Ni, Co: Dithizonates	Al_2O_3-thin film	0.25 – 5 ng	[241]
Zn, Be, La, Ba, Sn, Cd, Sr, Zr, Al: Oxinates	Polyamide	0.025 – 1 μg	[406]
Ni, Co, Cu, Fe, Pd, Pt, Ru: Diphenylthiovioluric acid chelates	TLC/A	0.04 – 0.8 μg	[190]

* See also the example of Se: *W. Bertsch, S. Hara, R. E. Kaiser, A. Zlatkis,* Instrumental HPTLC, published in the series "Chromatographic Methods". Dr. A. Hüthig Verlag, Heidelberg, New York (1980).

Table 35 (Continued)

1) Visual methods
b) after spraying

Elements: Spray reagent	Chromatography	Detection limit	Ref.
Sn: 2,2′-Diquinoxalyl	TLC/A	20 μg	[15]
Cu: Diethyldithiocarbamate	PC	40 ng	[225]
Ru, Os: Thiourea	PC	ng range	[334]
Cu, Co, Ni, Fe: Rubeanic acid	TLC/Al_2O_3	5 – 10 ng	[413]
Zn, Cu, Ni, Pb, Hg, Cd: as diethyldithiocarbamates + dithizone	TLC/A	ng range	[398]
Au, Se, Te: $SnCl_2$	TLC/Al_2O_3	0.5 μg	[124]
Nb: Tannin	PC	50 ng	[96]
Co, Ni, Cu, Bi, **Pb**, Mn, Cd, V, U: + TAR, PAR, PAN	TLC/C, S	10 ng	[289]
Co, Cu, Ni: + TAR	TLC	10 – 20 ng	[107]
As, Sb, Sn: Pyrrolidinedithiocarbamates + $CuSO_4$	TLC/A	100 – 300 ng	[350]
Halides: Bromocresol violet	TLC/P	1 – 2 μg	[413]
15 cations: survey	PC	0.2 – 0.03 μg	[383]

1) Visual methods
c) Fluorescence, fluorescence quenching

Elements: Reagent	Chromatography	Detection limit	Ref.
U: + Na_3PO_4/Zn acetate (intrinsic fluorescence)	PC	1 ng	[140]
Mg, Ca, Sr: 8-Hydroxyquinoline	PC	μg range	[255]
Be, Mg, Ca, Sr, Ba: 8-Hydroxyquinoline	TLC/C	0.1 – 0.5 μg	[119]

Table 35 (Continued)

Elements: Reagent	Chromatography	Detection limit	Ref.
Sn, Sb, Bi, Pb: + 2,6-Dibromoquinone chloroimide (fluorescence quenching)	TLC-$S_{UV_{254}}$	0.01 µg	[118]
Zn, Ag, Cd, Hg, Pb, Bi: + Pd-calcein (ligand exchange)	TLC/A as carbamates	1 – 2 ng	[65]

2) Planimetry
a) for colored substances

Elements: Compounds	Chromatography	Detection limits	Ref.
Cu, Co: 2-Hydrazinethiazole chelates	PC	0.1 – 6 µg	[233]
Mn, Co, Ni: 2-Hydrazinethiazole derivatives	PC	8,20,60 ng	[28]
S (as sulfate): as barium sulfate	PC (+ Ba rhodizonate)	0.22 µg	[25]

2) Planimetry
b) after spraying

Elements: Reagent	Chromatography	Detection limit	Ref.
Te: Bismuthiol II	PC	0.15 µg	[314]
Co: 1-Nitroso-2-naphthol	PC	0.31 µg	[218]
Re: $SnCl_2$/KSCN	TLC/ Al_2O_3-P	50 ng/µl sample	[123]
Pt, Pd: (Paper impregnated with metallic Ag)	PC	0.5 – 2 ng	[38]
Cd, Bi, Pb: 2-Mercapto-5-aniline-1,3,4-thiodiazole	PC	30.7 – 0.1 – 0.4 µg	[323]

Table 35 (Continued)

3) Spectrophotometry (transmission mode)
a) colored substances

Elements: Compound	Chromato-graphy	Detection limit	Ref.
S: + 2,2′-Dimethoxy-propane/HCl	TLC	0.3 µg (295 nm)	[315]
Ag: Dithizonate	TLC/A	(50 ppm) – 400 nm	[277]
Cd, Co, Ni, Zn, Cd: Dithizonates	TLC/A	1 µg range 490 – 538 nm	[90]

3) Spectrophotometry (transmission mode)
b) after spraying

Elements: Reagent	Chromato-graphy	Detection limit	Ref.
Se: Thiourea	PC	3 µg	[296]
Zn: PAN	TLC/C + liq. IE	1 µg	[138]
In: Alizarin	PC	0.25 µg	[208]
Li: U/Zn acetate (fluorescence)	TLC/C	ppm range	[276]
Ni, Co, Cu: Rubeanic acid	PC	2 µg	[163, 164]
Zn, Cd, Co, Fe: PAN	TLC/C + liq. IE	1 µg	[139]
Ni, Co, Cu, etc: miscellaneous	TLC/S	0.1 – 3.2 µg	[405]

4) Reflectance mode
a) for colored substances

Elements: Compounds	Chromato-graphy	Detection limit	Ref.
Cd: Dithizonate	TLC/S	µg range	[279]
Mn: PAN complex	TLC	(0.5 ppm)	[278]
Co, Cu, Ni: dithizonates	TLC/Al_2O_3	0.8 – 6.2 µg/g Al_2O_3	[230]

Table 35 (Continued)

4) Reflectance mode
b) after spraying

Elements: Reagent	Chromatography	Detection limit	Ref.
Co, Cu, Ni: TAR	TLC	1 ng	[107]
Ni, Co, Cu: Rubeanic acid	TLC	50 ng	[108]
Co, Ni, Zn: Neocuproin, diacetyldioxime	TLC/C	µg range	[116]
Co, Ni, Cu: Pyridin-2-aldehyde 2-quinolinylhydrazone	TLC/C, Al_2O_3	10 ng	[111]
Cu, Zn, Fe, Mn: miscellaneous	PC	1 µg	[436]
Co, Cu, Ni, Mo, Mn: miscellaneous	PC	50 ng	[75]

5) Fluorometry in situ

Elements: Reagent/compound	Chromatography	Detection limit	Ref.
Li: Zn/U acetate/-	TLC/C	ng-µg range	[276]
Pb: -/Chloro complex	TLC/C	0.3 µg	[418]
Sn: -/Chloro complex	TLC/C	0.5 µg	[419]
Alkaline earths: 8-Hydroxyquinoline/-	TLC/C	Ba, 50 ng; Sr, 20 ng; Ca, 10 ng; Mg, 5 ng	[280]

6) Electrometry in situ

Elements	Chromatography	Detection limit	Ref.
Alkaline earths	PC	50 µg	[71]

Table 35 (Continued)

7) Radiometry in situ

Element: Isotope	Chromatography	Detection limit	Ref.
Ni: $Ni^{35}S$	PC	10 ng	[72]
Na, K: ^{24}Na, ^{42}K	TLC	µg range	[343]
Sulfate, phosphate: ^{35}S, ^{32}P	In_2O_3: thin film	10^{-14} g	[80]

8) X-ray fluorescence in situ

Elements	Chromatography	Detection limit	Ref.
Cl, Br	TLC	20 – 2.5 µg	[172]

9) Determination after elution

Method	Elements: Reagent	Detection limit	Ref.
a) Photometry	Ni: MDCM (Carbamate)	µg range	[121]
	Co: MDCM (Carbamate)	µg range	[122]
	Bi: MDCM (Carbamate)	µg range	[120]
	Cu, Pb, Bi: 8-Hydroxyquinoline	4 ppm	[19]
	Cd, Co, Ni, Zn, Hg: Dithizonates	1 µg	[90]
b) Fluorometry	Al, Be: Salicylaldehyde	µg range	[23]
c) Emission spectrometry (flame photometry)	Alkaline earths	(0.05 ppm)	[109]
d) Infrared spectrometry	Co, Cu: Tetramethylenedithiocarbamate	5 µg	[214]
	Fe, Zn, Bi, Co, Cd, Mn, Pb, Hg, Cu: vide supra	µg range	[267]

Table 35 (Continued)

Method	Elements: Reagent	Detection limit	Ref.
e) Activation analysis	U	20 ng	[254]
	Cu, Zn	ppm range	[374]
	Cl-containing anions	ng range	[345]
f) Titrimetry	Zn: complexom. + Na arsenate/ Eriochrome black	3 µg	[337]
g) Ring oven colorimetry	Au, Ag, Ru: Thiocarbonate	1.6; 1; 1.2 µg	[187]
	Co, Cu, Ni, Fe: Rubeanic acid KSCN	5 – 10 ng	[110]
	Fe, V, Ti, Ge, As, Sb: various	0.08 – 3.82 µg	[188]
	Ni, Co, Pd, Ti, Y, Zr: various	90 – 300 ng	[285]
	Pb, Bi, Cu, Th, Ni, Co, Cd: various	µg range	[186]

10) Determination after digestion (for photometric determinations)

Element	Reagent	Detection limit	Ref.
Pb	PAR	µg range	[35]
Tl	Methyl violet	µg range	[37]
V	PAR	10 µg	[36]
Zn	8-Hydroxyquinoline	µg/ml serum	[66]
Chromatography:	PC – digestion with various acids		

Detailed studies on the semiquantitative determination of cations and anions with the aid of circular thin-layer chromatography by comparison with test solutions after spraying of the chromatograms have been published by *Hashmi* et al. [150 – 152].

The influence of various adsorption materials on the semiquantitative visual detection of toxic metals has been studied, for example, by *Thielemann,* 1975 [412].

Table 36

Reproducibility of various quantitative methods in layer chromatography of inorganic substances.
N, number of determinations; DL, detection limit; s, relative standard deviation; C, cellulose; S, silica gel; Ch, chromatography; IE, ion exchanger.

Method	Element: Compound, reagent	Ch: DL, s	Ref.
1b) *Visual*	Cu, Co, Ni: + TAR	TLC/C, S: 10 – 20 ng per element – 1.5%	[107]
1b) *Visual*	Ni: + Zn diethyl-xanthogenate	PC: 2 μg – 2%	[414]
2a) *Planimetry*	Sulfate: (C + Ba-rhodizonate impregnated)	PC: 0.22 to 1.33 μg s/spot – 0.6 to 1.5% (N = 6)	[25]
3a) *Densitometry* (295 nm)	S: (Reaction mit dimethoxypropane/HCl)	TLC/S: 0.3 μg – 5%	[315]
3b) *Densitometry*	Numerous elements: + various reagents	TLC/S: for 0.1 to 3.2 μg – 0.2 – 2.1%	[405]
4b) *Reflectance*	Co, Cu, Ni: + TAR 4-(2-thiazolylazo)resorcinol	TLC/S, C: DL = 1 ng; Co: 50 ng/10.27%, 0.2 μg/5.60%, 0.6 μg/3.43% (N = 9/9/6)	[107]
4b) *Reflectance*	Ni, Co, Cu: + Pyridin-2-aldehyde 2-quinolyl-hydrazone	TLC/S: DL = 10 ng/spot; for 1 μg: Co, 0.85%; Ni, 0.78%; Cu, 1.02%	[111]
4b) *Reflectance*	U: + $K_3Fe(CN)_6$	PC: 10 – 50 μg/dm^3 – 3%	[26]
4b) *Reflectance*	Zn, Cd, Co, Fe: + PAN	TLC/C + liq. IE: for 1 μg – 4%	[139]
4b) *Reflectance*	Zn: + PAN	TLC/C + liq. IE: for 1 μg – 5%	[138]
4b) *Reflectance*	Cu, Ni, Zn: Neocuproin, diacetyldioxime, 3,3-di-ethylnaphthidine	TLC/C: Cu – 16 μg/3.5%; Ni – 7 μg/4.28%; Zn – 26 μg/2.7%	[116]
4b) *Reflectance*	Ni, Co, Cu: + Rubeanic acid	TLC/Al, S, C: DL: 0.05 μg; 2 – 5%	[108]
4b) *Reflectance*	Co, Cu, Ni: Rubeanic acid	TLC: (N = 4) (DL in ppm) Ni: 0.98/6.4%; Cu: 3.48/4.9%; Co: 0.025/20%	[106]
5) *Fluorometry in situ*	Mg, Ca, Sr: + 8-Hydroxy-quinoline	PC: μg range 2%	[255]
Elution			
9a) *Photometry*	Numerous metals: Dithizonates	TLC/S: 0.125 to 0.5 ppm – 4%	[386]
9a)	Cu: Dithizonate	PC: 4 μg/sample – 1.3%	[371]

Table 36 (Continued)

Method	Element: Compound Reagen	Ch: DL, s	Ref.
9a) (800 nm)	P: Phosphate + molybdenum reaction	TLC/C: 0.2 μg P – 20 – 25%	[206]
9a) (365 nm)	Bi: Extraction as carbamate (MDCM)	TLC/C: for 2 – 40 μg Bi – 15 to 2.5%	[120]
9c) *Emission*	Ba, Ca, Mg, Sr: Elution as oxinates	TLC/C: 0.05 ppm – 10%	[109]
9e) IR	Co: Elution as carbamate (TMDTC)	TLC/S: 5 μg Co-TMDTC – 2.38% (N = 6)	[214]
9a)	Cu, Zn, Fe, Mn: various reagents	PC: μg range – Fe, 5.8%; Zn, 11.8%; Mn, 5.1%; Cu, 11.1%	[436]
10) *Digestion* AAS	Ag: – Digestion of 5 mg cellulose material	Chelate-cellulose (azorhodanine group): 10^{-6}% – 20% (s) 5×10^{-7}% – 33%	[297]
Miscellaneous			
2b) *Planimetry*	Na, K, Mg, and Cu, Co, Ni: + Violuric acid	TLC: μg range – 10%	
3b) *Transmission*	and rubeanic acid	TLC: μg range – 3 and 4%	
7) *Radiometry*	^{24}Na, ^{42}K	TLC: μg range – 1%	[343]

The fundamentals of photometric evaluation of paper chromatograms were reported by *Hejtmanek,* 1966 [163].

The **quantitative results** of different paper or thin-layer chromatographic techniques reported in various sources frequently do not permit direct comparison. Individual authors give widely differing information for quantitative characterization of the methods described.

Apart from the **detection limit**, which is also not defined uniformly, we find data pertaining to the **determination range** and the **deviation from the true amount or error**. The relative standard deviation (coefficient of variation) refers either to the detection limit or, more commonly, to a particular analytical result. The detection limit itself is given either as the absolute amount in the layer, and less frequently as the amount per gram of layer material, as the concentration in the applied sample, or in the case of a practically oriented technique, as concentration in the sample material.

The widely differing data make any comparison of techniques difficult or even impossible. Nevertheless, wherever they are given in the original literature, the detection limits and detection ranges have been compiled in Table 35 in order to convey some idea of the power and limitations of quantitative layer chromatography. Furthermore, Table 36 gives detailed information from selected publications about the **reproducibility** of various methods of determination.

2 Column Chromatography

In classical column chromatography, fractions of the eluate were investigated by a wide variety of methods of inorganic analysis to establish their content of elements. As a consequence of the relatively large volumes required for separation of ions or also compounds, considerable dilution effects often occurred, so that only sensitive methods came into question for quantitative trace analysis in the eluate (without additional enrichment). Table 37 presents a collection of such combined procedures from column chromatography and various determination methods, in which chromatography serves for removal of interfering substances (or even the matrix), for separation into fractions of the individual elements, or for enrichment from large volumes.

2.1 Spectroscopic Detectors

Column chromatography is presently undergoing a renaissance as high performance liquid chromatography (HPLC) – not only because of improved separation performance but also owing to greatly improved possibilities of detection by on-line coupling with specially developed detectors. The advantages of sensitive detection coupled directly with the separation process for several substances are, however, somewhat offset by the drawbacks, in say UV/VIS photometry, that (i) the eluted substances may exhibit different absorption maxima while measurement is possible only at one wavelength with most detectors and (ii) the required (UV) transparency of the mobile phase limits the scope of both detection and chromatography. On the other, precisely these limitations may lead to selectivity effects in special cases.

Table 38 surveys other spectroscopic detectors such as flame photometers, atomic absorption photometers, fluorometers, and microwave emission detectors which have also been used on occasion in direct conjunction with liquid chromatography.

Table 37

Selection of methods of determination in combination with method of column chromatography.

PS, pre-separation (removal of interfering substances); MT, removal of matrix; E, enrichment; P, partition; C, cellulose; IE, ion exchange; RP, reversed phase.

Method of determination	Elements	Chromatography	Detection limit	Ref.
Volumetry	Sn	P-C	μg range	[99]
	Cr, Cu, Pb, Fe	E/IE	10^{-6} mol/dm^3	[312]
	Pb	MT/IE	ppm range	[400]
Photometry	B	E/IE	μg range	[264]
	Sn	MT/C	80 ppm	[2]
	Sn	PS/IE	0.1 ppm	[369]
	Cr, V	E/Chelate IE	μg amounts	[283]
	Zr, Hf	PS/RP	–	[74]
	Pt group	PS/IE	ppm-ppb range	[346]
	Pb, U, Th	MT/IE	ppm range	[400]
	Zn, Cd, Hg	MT/modif. SiO_2	ng amounts	[262]
Fluorometry	Ce	E/RP	1 ng/dm^3	[359]
	In	MT/RP	18 ppm	[182]
	Tl	PS/IE	0.1 ppm	[282]
	Element traces	PS, E/IE	ppb range	[227]
AAS	Ag	E/IE	0.1 ppb	[70]
	Ba	E/IE	10 ppm	[365]
	Be	MT/IE	ppb range	[228]
	Bi	MT/RP	ppm range	[381]
	Element traces	PS, E/IE	ppb range	[227]
	Metals	PS/IE	ppm range	[229]
	Cu, Ni, Zn, Cd	PS/Chelate IE	ppm range	[421]
	Pb, Cd	PS/IE	1 ppm	[431]
	Na, K, Mg, Ca	MT/IE gel	1 – 20 ppm	[231]
Emission spectrometry	Alkalis, alkaline earths	MT/Sephadex gel	ppm range	[205]
	Rare earths	MT/C	ng range	[199]
X-ray fluorescence	Element traces	E/Chelate IE	1 ppb	[58]
	U	E/Chelate IE	0.3 ppb	[59]
Neutron activation	Tl	PS/IE	0.1 ppm	[282]
	U	PS/RP	0.5 ng	[437]
	Metals	MT/IE	10^{-5} – 10^{-10}%	[302]
	Cu, Zn	PS/IE	ppm range	[374]
	V, Al	MT/IE	0.15 ppm	[301]
Voltammetry	Cu, Cd, Pb	MT/IE	10 ppm	

Table 38
Spectroscopic detection for liquid chromatography of inorganic substances on columns.

Compound: Elements	Wave-length, nm	Detection limit	Ref.
a) *UV/VIS detection*			
Schiff base			
Chelates: Ni, Cu	254	25 ng per chelate	[422]
Tetradentate β-keto amine complexes: Ni, Cu	254	0.2 ng Ni, 0.5 ng Cu abs.	[117a]
Diacetyl-bis-thiobenzhydrazone compl.: Hg, Cu	360	1 ng Hg, Cu abs.	[162]
Dithizonates: Hg, Cu	525	10 ng Hg, Cu abs.	[261]
Diethyldithiocarbamates: Co, Ni, Cu	254	5 – 10 ng abs.	[423]
5-Nitropiazselenole: Se	340	0.5 ng Se abs.	[392]
5-Chloropiazselenole: Se	320	0.32 ppb (sample)	[399]
Sulfur, elemental	263	30 ng abs.	[291]
Sulfur, elemental	254	1 ng abs.	[64]
Nitrate	210	0.25 mg NO_3^-- nitrogen	[204, 130]
Ion pairs with cetyltrimethyl-ammonium: anions (as K salts)	215	IO_3^-: 5 ng, NO_2^-: 1.2 ng, NO_3^-: 10 ng, I^-: 3 ng, $S_4O_6^{2-}$: 5 ng etc.	[335]
b) *Flame photometer*			
Ions of rare earths, alkalis and alkaline earths		for Ca: 2 ppm, for Na: 0.1 ppm	[105]
NTA-, EDTA complexes: Cu		0.01 mg/dm^3	[197]
c) *Atomic absorption spectrometer (AAS)*			
Organometallic cpds.: Hg		10 ng Ethyl-Hg-Cl	[117]
Organometallic cpds.: As, Pb, Hg, Sn		between 4.8 and 111 ng (flameless AAS)	[54]
NTA-, EDTA complexes: Cu		3.13 ng Cu-NTA, 6.72 ng Cu-EDTA	[197]
NTA-, EDTA complexes: Cu		13.5 – 450 ng Cu abs.	[196]
Organomet. complexes: Cr		4 μg/cm^3 mobile phase	[195]
Amino acid complexes: Cu		(flameless AAS) –	[202]
Organomet. complexes: Pb		ng range	[224]
Condensed phosphates of Mg		–	[226, 450]
d) *Fluorometer*			
Nitrates of rare earths		– (fluorescence quenching)	[439]
Naphthylpiazselenole: Se, 2,3-naphthotriazole: NO_2^-		10 ng/cm^3	[440]
e) *Microwave emission detector*			
Diethyldithiocarbamates: Co, Cu, Ni		–	[423]

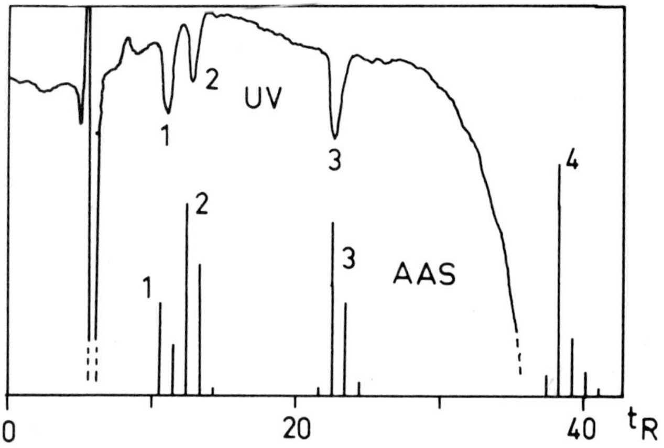

Figure 32
HPLC separation of organomercurials – comparison of UV and AAS detection according to *Brinkman* et al., 1977 [54].
1, n-$C_3H_7Hg^+$ (220 ng); 2, CH_3Hg^+ (250 ng); 3, $C_2H_5Hg^+$ (221 ng); 4, $C_6H_5Hg^+$ (218 ng).
Packing: LiChrosorb RP-8.
Mobile phase: A, 0.01 M ammonium acetate in water; B, 25 ppm mecaptoethanol in methanol; 25 min isocratic development (96% A + 4% B), then gradient to 100% B at 10% min^{-1}.
Flow rate: 0.3 cm^3 min (t_R in min).
Pressure: from 34 to 29 bar.
Detection: UV at 254 nm, AAS – flameless (with automatic fraction collector).

2.2 Electrochemical Detectors

Various electrochemical detectors such as conductometers, potentiometers, coulometers, and polarographs has so far been employed only rarely as flow-through detectors in column chromatography. This is due, on the one hand, to the fact that electrochemical methods can often be used (as in water analysis) without a chromatographic separation step for simultaneous determination of several elements without mutual interference in a single operation (for example by methods based on voltammetry) and, on the other, to the instrumental problems posed especially in continual operation of the measuring cells by the large volumes of mobile phase (usually salt solutions). The success of ion chromatography (see Section I/1.2.4) has made conductometry particularly interesting for the detection of ions. A very recent

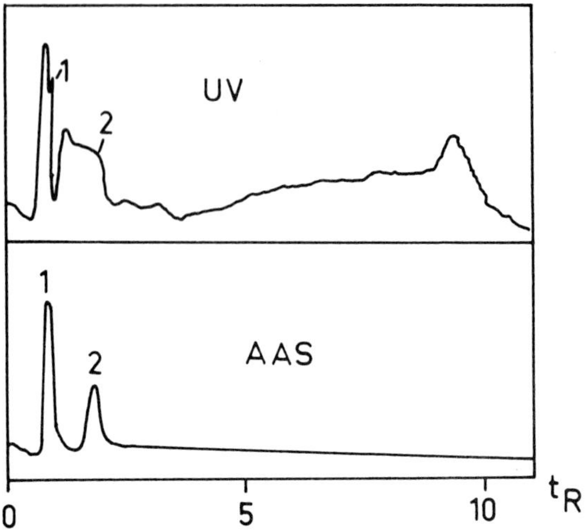

Figure 33

Ion exchange chromatography of copper chelates in water – comparison of UV and AAS detection according to *Jones* and *Manahan*, 1976 [196].
1, Cu – NTA$^-$; 2, Cu – EDTA^{2-}.
Packing: Aminex A-14/30 μm ion exchanger, column dimensions 50 × 2.1 mm.
Mobile phase: 0.05 ammonium sulfate in water.
Flow rate: 2.0 cm^3 min^{-1} (t_R in min).
UV detector: at 235 nm.
AAS detector: flame AAS, on-line coupling.

article by *Girard*, 1977 [133] is concerted with the possibilities of coulometric detection in ion chromatography.

The application of these detectors is limited to ion chromatography and thus complements that of spectroscopic dectors for determination of organometallic compounds and metal chelates.

2.3 Chemical Reaction Detectors

Detection systems in which chemical reactions occur immediately after leaving the column and prior to the actual detection are designated as "(chemical) reaction detectors" (see *Schwedt*, 1979 [397a]). In the simplest case the reagent is introduced directly into the column effluent from a capillary tube (open reaction tube), while for multistep reactions or long reaction times an air-segmented mixing coil reaction

Table 39

Electrochemical flow-through detectors for inorganic column chromatography.

Ions/Applications	Method	Ref.
Anions and cations	Conductometry	[367]
Halides	Potentiometry (Ag/AgCl-electrode)	[104]
Nitrate/nitrite	Coulometry: Cd electrode	[81]
Sb(III) cation	Coulometry: irreversible electrochemical oxidation at a Pt electrode	[409]
Inorganic ions: Ion chromatography	Coulometry	[133]
Chromium(VI): Chromate	Coulometry: Reduction at a Pt electrode (electrocatalyzed by adsorbed iodine)	[240]
Cations	Polarography	[102]
For HPLC	Voltammetry: dropping Hg electrode	[288]
Hg-organic species	Voltammetry	[265]

unit may be used (Auto-Analyzer type). Various metal reagents have recently been used in such reaction detectors for ion-exchange chromatography. The literature is surveyed in Table 40.

The use of metal reagents known from photometric analysis permits quantitative analyses in photometric reaction detectors after separation on ion exchangers. Owing to the very short reaction times, an open reaction tube is usually suitable as reactor.

However, an air-segmented mixing coil reaction unit has been described by *Matsui* [281] for determination of **zinc** in water samples. Traces of zinc can be enriched from volumes of up to two liters on Amberlite CG-120 ion exchanger, giving a detection limit of 2 µg in water or 2 ppb in the column. Zincon served as reagent.

The comparatively unspecific metal reagents PAR and PAN have been used in the reactor for multielement analysis. Separations were performed with macroreticular strongly acidic cation exchangers. A

Table 40
Ion chromatography with reaction detectors.

Elements	Reagent	Detection limit	Ref.
A) *Cations*			
Zn	Zincon	2 μg abs.	[281]
Zn, Cd, Pb	PAR	$1-5 \times 10^{-5}$ mol/dm^3	[319]
Cu, Co	Luminol + H_2O_2	10^{-7} mole	[300]
Cd, Zn, Fe, Pb, Cu, Co, Mn	PAR, PAN	$10^{-5}-10^{-8}$ mole abs.	[211–213]
Cu, Co, Mn, V, Ni	PAR	μg range	[210]
13 cations	PAR-ZnEDTA	$10^{-9}-10^{-10}$ mole abs.	[184]
20 metal ions	Luminol + H_2O_2	–	[149]
Lanthanides	PAR, Arsenazo I	0.1 – 1 ng abs.	[92]
B) *Anions*			
Polythionates	Ce(IV) sulfate	0.3 ppm (in mobile phase)	[442, 443]
PO_3F^{2-}	Molybdate/hydrazine	(0.1% in toothpaste)	[24]
Polyphosphates	Molybdate/ascorbic acid	10^{-7} mole	[166]
Linear and cyclic condensed phosphates	Molybdate reaction	μmole range	[444]

stepwise gradient system has been described by *Kawazu* and co-workers [210 – 213] for separation of a larger number of cations: the mobile phase consists of hydrochloric acid, water, and acetone. The concentration of hydrochloric acid is increased from 0.1 to 1.88 mol/dm^3, and the acetone content from 70 to 96%. This system permits separation of a large number of cations and their detection in the range between 10^{-5} and 10^{-8} mole.

Practical applications, for example in water analysis, have not yet been described for this method. With Amberlite 200 (25 – 30 μm), stepwise elution by hydrochloric acid/acetone and hydrochloric acid/acetone/dimethylformamide mixtures can separate 11 cations which can be detected in the 10^{-8} mole range at 525 nm after reaction with PAR [211]. Pressures of 16 – 25 bar are used for a column of dimensions 120 × 5 mm and a separation temperature of 40 °C; the analysis time is about 35 min.

The sensitivity of detection can be enhanced with PAR-ZnEDTA as reagent [184]. Detection limits between 0.1 and 1.0 nmol are reported for 13 metal ions; detection is at 510 nm. Separations of individual metal ions on immobilized 8-hydroxyquinoline materials have been described with sodium sulfate/ sodium hydrogen sulfate solutions of various pH values being used as mobile phase. Separation of all metal ions investigated with a single system has not yet been possible.

Chemically bonded ion exchange materials with particle diameters of 5 to 10 μm are suitable for fast separation of lanthanides in less than 20 min with mobile phases containing hydroxybutyric acid [92]. Elution is performed with a linear program of 0.018 to 0.070 mol/dm^3 (pH 4.6). Detection limits of 1.0 to 1 ng, depending upon the lanthanide, can be achieved with PAR and arsenazo I as reagents, and quantitative analyses are feasible in the range of 10 to 600 ng per element in the column. The maxima of the metal complexes lie at 490 nm (PAR) and 585 nm (arsenazo I). Various packings (chemically bonded and conventional ion exchangers) have been compared with regard to their performance.

Metal-ion catalyzed oxidation of luminol by hydrogen peroxide provides a highly sensitive means of detecting these ions by means of a **chemiluminescence detector** [300]. The chemiluminescence can be recorded by a photomultiplier; a complete fluorometer is not necessary. However, numerous possible sources of interference in practical applications suggest that the method may be of limited practicability.

By way of contrast, another reaction amenable to fluorometry, viz. oxidation of polythionate ions by cerium(IV) ions in which the resulting cerium(III) ions fluoresce (excitation at 260 nm, emission at 350 nm) has already been used for quantitative analysis of these anions in waste water. Figure 34b compares UV detection at 254 nm and fluorescence detection.

The determination of **monofluorophosphate ions** in toothpaste, after separation from orthophosphate ions, by conventional ion-exchange chromatography was described as early as 1969 [24]. The reaction with

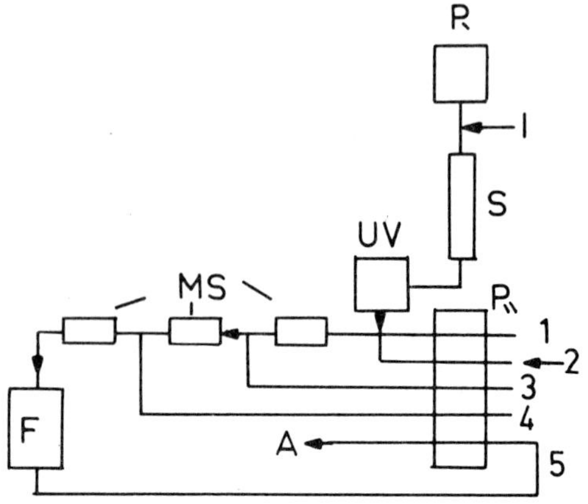

Figure 34a

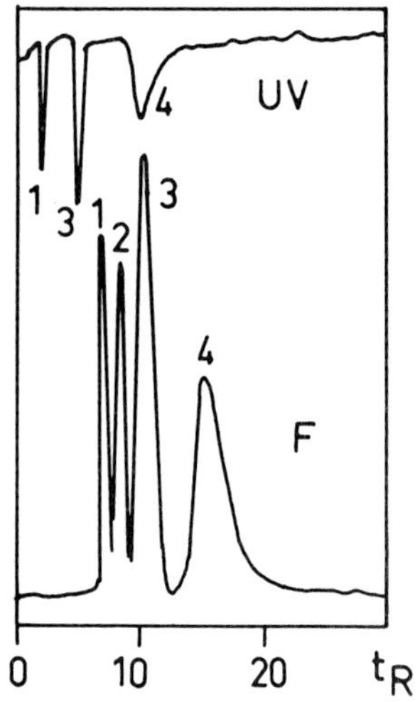

Figure 34b

Figure 34a

Scheme of a reaction detector for HPLC of polythionates, according to *Wolkoff* and *Larose*, 1975 [442].

P_I, HPLC pump; I, sample injection device; S, HPLC column; UV, UV flow-through detector; P_{II}, perstaltic pump of reaction detector; 1 – 5, tubing supplying (1) air (for segmentation of liquid flow), (2) NaOH, (3) H_2SO_4, (4) $Ce(SO_4)_2$, (5) overall supply tube (feeding overall volume through the fluorometer F); A, discharge of measured solution; MS, glass mixing coils.

Figure 34b

HPLC of thiosulfate and polythionate ions with UV and fluorometric reaction detector, according to *Wolkoff* and *Larose*, 1975 [442].

Packing: Permaphase AAX anion exchanger, column dimensions 1000 × 2.1 mm.

Mobile phase: 10^{-3}M aqueous citrate solution.

Flow rate: 1.0 cm^3 min^{-1} (t_R in min).

Detection: UV at 254 nm; fluorometry (F) at 260/350 nm.

Sample: 1, 12.8 ppm $S_2O_3^{2-}$; 2, 40.0 ppm $S_3O_6^{2-}$; 3, 45.0 ppm $S_4O_6^{2-}$; 30.6 ppm $S_5O_6^{2-}$.

molybdate and hydrazine to give molybdenum blue (measurement of absorption at 660 nm) was performed in a "Technicon AutoAnalyzer". Hydrolysis of monofluorophosphate at 95 °C and dialysis through standard membranes at 38 °C to remove interfering components precede the actual chemical reaction.

An HPLC method for separation of ortho-, di-, and triphosphates on an anion exchanger in combination with a reaction detector has been described by *Hirai, Yoza*, and *Ohashi* [166]. The authors use a mixing coil reaction unit. Hydrolysis is performed at 98 °C, reaction with molybdate at 82 °C; reduction is with ascorbic acid. The molybdenum blue formed is measured in a flow-through photometer at 830 nm.

The use of reaction detectors in chromatography of inorganic ions often facilitates sample preparation and can also be utilized in cases of incomplete separation, thanks to the selectivity of the chemical reactions (or reagents). Chemical reaction detectors therefore compete with electrochemical detectors.

2.4 Special Detectors

Two special detectors for liquid chromatography in columns, which are also employed in gas chromatography, are the **flame ionization detector** (FID) and the **radiometric detector**.

The FID has two flames serving for evaporation and for subsequent ionization of the mobile phase and the eluted substance, respectively.

Both detectors are used exclusively in ion-exchange chromatography of **alkali** and **alkaline earth metals**. With respect to sensitivity, the FID is superior to the radiometric flow-through detector (with limited counting times).

Table 41
Special detectors for liquid column chromatography.

Elements	Detection limits	Ref.
a) Flame ionization detector (double FID)		
K, Rb, Cs	10^{-14} mole	[10]
Na, Sr, Ba	10^{-13} mole	[11]
Li, Ca	10^{-12} mole	[11]
b) Radiometric detector		
Na, K, Rb, Cs	–	[174]
Na, K, Rb, Cs	0.07, 0.2, 3.5, 6.5 µg abs.	[180]
Alkalis	µg-mg range	[32]

3 Gas Chromatography

3.1 Conventional Detectors: TCD, FID, ECD

The **thermal conductivity detector** (TCD) is suitable for both inorganic and organometallic compounds, as well as for the many metal chelates and other organic derivatives; however, it has a relatively low sensitivity of 10^{-4} to 10^{-8} g, with respect to the metal, and no selectivity. In this context we use the term **selectivity** to mean the ratio of the amount of substance to be determined to the amount of purely organic compounds which give rise to the same detector signals. Detection of metal chlorides requires use of special corrosion-resistant cells.

The **flame-ionization detector** (FID) responds to substances having CH_2 groups. The detection limits for organometallic compounds are at least three orders of magnitude lower that with the TCD (around 10^{-11} g). Certain combustion gas compositions can lead to selectivity effects, e.g. for Si compounds [587]; particularly low detection limits can be attained for some organometallic substances on use of combustion gases with high hydrogen contents [464].

Systematic studies on the detection of purely inorganic compounds with a FID have shown, however, that some of these substances, especially oxygen-containing gases, can be detected under certain conditions (see Table 42). In contrast to the TCD, the substances detected are destroyed in the FID.

In spite of its smaller linear response range compared to the TCD and FID, the **electron capture detector** (ECD) enjoys the widest application in quantitative analysis of metal chelates. Its sensitivity is two to three orders of magnitude higher again, in the most favorable case, than the FID. It shows a pronounced selectivity for halogenated substances, and is therefore used primarily for detection of fluorinated metal chelates, for which detection limits of 10^{-13} mol are attained.

Furthermore, the ECD can also be used for sensitive determination of halogen-containing inorganic compounds, such as metal chlorides, and sulfur-containing compounds (see Table 43). Compounds containing nitro groups, such as 5-nitropiazselenole, can likewise be sensitively detected by the ECD, thus permitting selenium determinations in the picogram range.

Table 42

Detection of inorganic compounds with a FID.

Element: Compound	Detection limit	Ref.
N_2, O_2	–	[571]
O_2, NO_2	–	[742]
H_2O	–	[743]
CO, CO_2	–	[744]
H_2S, SO_2	–	[741]
Refs. [741 – 744]: general studies		
Fe: Ferrocene	*10 pg	[464]
Pb: Tetraethyl-	*30pg	[464]
Sn: Tetraethyl-	*20 pg	[464]
Si: Organic cpds.	**ng range	[587]

* H_2-rich flame, ** modified FID.

Table 43

Detection of inorganic compounds with an ECD – examples.

Element: Compound	Detection limit	Ref.
Cr: Trifluoroacetylacetonate	4.1×10^{-13} g	[461]
Si: $SiCl_4$	2.9×10^{-11} mole	[719]
Ge: $GeCl_4$	6.5×10^{-9} mole	[719]
P: PCl_3	1.7×10^{-11} mole	[719]
P: $POCl_3$	1.6×10^{-10} mole	[719]
P: $PSCl_3$	5.9×10^{-10} mole	[719]
H_2S	500 ppm	[618]
COS	100 ppm	[618]
SO_2	5 ppm	[618]
Se: Subst. (dichloro-)piazselenoles	10 pg	[758]
Se: Various piazselenoles	pg range	[726]
Nitrate as nitrobenzene	1 pg	[693]

The physical and instrumental fundamentals of these detectors have been considered in detail in the monographs by *Sevcik* [719] and by *Jentzsch* and *Otte* [604 a].

For liquid and solid substances the detection limits refer to the amount injected which can be distinguished from detector noise. However, the quantitative definitions are different. For gases the detection limit is given in ppm referred to the carrier gas, and referred to the analytical sample in practically oriented work.

3.2 Special Ionization Detectors

The **thermionic detector** (TID) is based on the FID: an alkali metal salt is located in the upper flame of a double FID, and its enhanced emission on combustion of halogen- and metal-containing substances is measured. In spite of its selectivity for some elements (e.g. N, P, As), the low sensitivity and susceptibility to interference hamper wide use of this detector in inorganic analysis (see *Sevcik* [719]). Two recent publications are concerned, for example, with the determination of phosphorus as its hydride [489] and of ammonia [679].

Table 44

Applications of helium and argon detectors in gas chromatography of inorganic substances.

Compound	Detection limit	Ref.
Helium detector		
N_2, O_2	0.1 mm^3 abs.	[564]
CO	50 ppb	[504]
H_2, O_2, Ar, N_2, CO, CO_2	1 ppb and below	[575]
Argon detector		
CO_2, NO_2	ppm range	[589]
H_2S, SO_2	10^{-3}, 5×10^{-3}%	[812]
H_2	10^{-10} g/s	[558]
He, Kr	7.4×10^{-8}, 1.56×10^{-6} g/s	[501]
H_2, O_2, N_2	3.76×10^{-10}, 2.98×10^{-9}, 3.93×10^{-9} g/s	[501]
CO, CO_2	2.61×10^{-9}, 4.15×10^{-10} g/s	[501]

Ionization detectors which, as variants of the ECD, use helium or argon as carrier gas in place of nitrogen, are known respectively as **helium** and **argon detectors** and are used primarily in the analysis of permanent gases. The principle of these detectors is considered in detail in the monograph by *Sevcik*. A selection of applications has been compiled in Table 44.

3.3 Photometric Detectors

The **flame photometric detector** (FPD) has roughly the same, or sometimes higher, sensitivity as the FID, and a high element selectivity since the wavelength can be adjusted to an optimum value where there is no interference from other substances in the gas chromatogram of metal-containing compounds. Characteristic emission wavelengths of various elements are listed in Table 45.

During in 1970's, the FPD was widely used in gas chromatography of a series of elements as inorganic or organic compounds. Table 46 surveys the compounds detected, and their detection limits, arranged according to elements. The lowest detection limit was found for tin compounds, at 10^{-12} g (= 1 pg) of Sn [463]. This detector has been successfully employed for numerous analyses of environmental samples (see Section III/2.4).

During the past few years, the FPD seems to have been increasingly supplanted by a detector of comparable selectivity, i.e. **the microwave emission detector** MED, in spite of considerably greater demands with respect to instrumentation. Since the mid-1970's, preference has been given to the so-called inductively-coupled plasma detector (ICP). Argon carrier gas is passed from the column into a quartz capillary located at the center of a cavity resonator. The intensity of the most favorable emission line of an element is measured with the aid of a photomultiplier. Quantitative analyses of many elements are possible in the picogram range. The fundamental problems of this detector have been considered, for example, by *Talmi* and *Andres,* 1974 [762], *Serravallo* and *Risby,* 1975 [718], and *Windsor,* 1979 [800].

Less use has so far been made of the **atomic absorption spectrophotometer** (AAS), which is likewise a highly selective detector for inorganic analysis. The instrumental requirements for coupling an AAS device to a gas chromatograph have occupied, for example, *Seger,* 1974 [714] and *Wolff,* 1976/77 [802, 803]. Applications of both flame AAS and flameless AAS have been reported for the elements Cr, Hg, Sb, Se, and Pb in the form of various compounds (see Table 48). Since the utilization of AAS for detection in gas chromatography began only a few years ago (from about 1974), any statements regarding its general suitability would be premature. Its sensitivity has not yet surpassed that of MED.

Another spectroscopic detector widely used in organic analysis, viz. the **mass spectrometer** (MS), has found only very limited application to specific problems in GC analysis of inorganic substances, generally for identification rather than quantitative analysis. A survey is presented in Table 49. Use of the integrated ion current techniques give detection limits of, for example, 10^{-14} g (= 0.01 pg) for fluorinated Pb diketonates [478]. The selectivity for metal chelates

Table 45

Emission wavelengths of elements for analysis by gas chromatography with a flame photometric detector (FPD).

Element	Wavelength	Ref.
Al	486 nm	[809]
As	500.0 nm	[608]
B	520 nm; 546 nm	[586; 737]
Cr	425.4 nm; 427 nm	[687; 809]
Cu	327.4 nm	[586]
Fe	373.5 nm	in: [719]
Mo	520 nm	[809]
Ni	341.6 nm	[586]
P	526 nm	[567; 465; 586]
Pb	405.8 nm	[465]
Rh	369.2 nm	[608]
S	384 nm; 394 nm	in: [719]
Se	335 nm	[586]
Sn	358 nm; 485.0 nm	[809; 465]
Ti	544.9 nm	[608]
W	520 nm	[809]
Zr	564.0 nm	[608]

with the same ligands is, however, far lower than that of FPD, MED, and AAS.

Tables 45 – 49 summarize publications about the use of photometric (spectroscopic) detectors in gas chromatography of inorganic substances.

Table 46

Applications of the flame photometric detector (FPD) in gas chromatography of inorganic substances.

Element: Compound	Detection limit	Ref.
B: $B_{10}H_{14}$	0.71 ng	[737]
B: Methyl borate	1.2 μg	[586]
Cr: Hexafluoroacetyl-acetonate	1 μg	[588]
Trifluoroacetylacetonate	2.5 ng	[687]
Cu: Hexafluoroacetyl-acetonate	670 ng	[586]
Ni: Dicyclopentadienyl-	120 ng	[586]
P: PH_3	5 pg	[489]
P: Elemental P	40 pg; 0.2 ppb	[567; 454]
P: Triethyl phosphate	80 ng	[586]
S: SO_2, H_2S	10 ppb	[666]
S: SO_2	10 ppb	[527]
S: SO_2, H_2S, CH_3SH	1 ppb	[757]
S: SO_2	3 ng	[713]
S: H_2S, SO_2	10 ppm	[780]
S: H_2S, SO_2	(general information)	[566]
Se: Piazselenoles	600 ng	[586]
Se: Piazselenoles	2×10^{-10} g	[545]
Sn: $SnCl_4$	2×10^{-9} g	[809]
Sn: Sn organic cpds.	10^{-12}g	[463]
Pb: Tetraethyl-	40 ng	[586]
S, P: Various cpds.	10^{-10}, 10^{-12} g	[579]
Cr, Rh: Hexafluoroacetyl-acetonates	10^{-10}, 10^{-11} mole	[608]
Mo, W: Fluorides	2×10^{-11}, 6×10^{-11} g	[809]
Al, Cr, Rh: Acetyl-acetonates	2×10^{-10}, 2×10^{-10}, 7×10^{-10} g	[809]
P, Pb, Sn: Tri-, tetraethyl-	10^{-9}, 3×10^{-11}, 2×10^{-11} g	[465]

Table 47

Applications of the microwave emission detector (MED) in gas chromatography of inorganic substances.

Element: Compound	Detection limit	Ref.
As: AsH_3	0.2 ppm	[705]
As: Alkylarsanes	20 pg As	[763]
Cr: Trifluoroacetylacetonate	(general information)	[718]
Cr: Trifluoroacetylacetonate	9×10^{-13} g	[493]
Be: Trifluoroacetylacetonate	1 pg	[706]
Hg: Dimethylmercury	0.3 ng	[568]
Hg: Hg organic cpds.	1 pg Hg	[761]
Se: Nitropiazselenole	40 pg	[762]
Si: Alkylsilanes	1 ng and below	[500]
Pb: Tetraalkyl cpds.	17 ng/m^3 (air)	[680]
Al, Cu: Trifluoroacetyl-acetonates	1 ng Cu; 0.5 ng Al	[704]
Al, Sc: Trifluoroacetyl-acetonates	2×10^{-11}, 2.1×10^{-12} g	[529]
As, Sb: Triphenyl-	pg range	[764]
Al, Be, Cr, Cu, Fe, Ge, In: Trifluoroacetylacetonates	$10^{-10} - 10^{-7}$ g	[616]
As, Ge, Se, Sb, Sn: Hydrides	7, 3, 25, 10, 40 ng	[682]
Si, Fe, Pb, Sn: Tetraethyl-	0.8, 5.9, 33, 0.9 ng	[800]

Fundamental publications on MED:
[716 – 718], [529], [800]: ICP.

Table 48

Applications of the atomic absorption spectrophotometer (AAS) as detector in gas chromatography of inorganic substances.

Element: Compound	Detection limit	Ref.
Flameless AAS: general information		[714]
Cr: Trifluoroacetylacetonate	1 ng	[802, 803]
Hg: Alkyl cpds.	ng range	[516]
Pb: Alkyl cpds.	14 ng/m^3 Pb	[678]
Pb: Tetraethyl- (flameless AAS)	10^{-10} g	[683]
Sb: $SbCl_3$	50 ppm (flame), 3 ppm (flameless)	[695]
Se: Dimethyl selenide, dimethyl diselenide	10 ng Se	[521]

Table 49

Applications of the mass spectrometer (MS) as detector in gas chromatography of inorganic substances.

Element: Compound	Detection limit	Ref.
Cl: HCl, Cl_2, $SiCl_4$	For identification	[458]
$^{18}O/^{16}O$ ratio: Phosphates	–	[471]
Si: Trimethylsilyl esters	–	[456, 806]
Cu, Ni, etc.: Dialkyldithiocarbamates	–	[630]
Cu, Fe, Rh, Cr: Acetylacetonates	(general information)	[766]
Cr, Be: Trifluoroacetylacetonates	10^{-12} g	[805]
Pb: Fluorinated β-diketonates (integration of ion current)	10^{-14} g (m/e: 503)	[478]

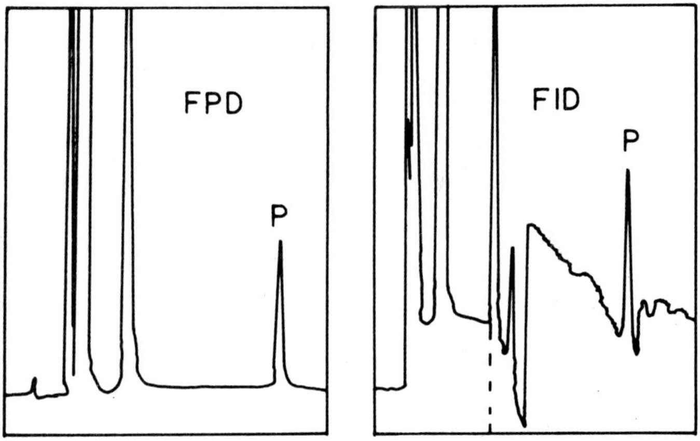

Figure 35

Detection of triethylsilyl phosphate with a flame photometric (FPD) and a flame ionization GC detector (FID), according to *Matthews* et al., 1971 [586].
Packing: 5% OV-225 on Chromosorb W(HP) – 80/100 mesh, column dimensions 1830 × 6 mm.
Column temperature: programmed from 65 °C at 5 °C/min.
Carrier gas: nitrogen.
Flow rate: 80 cm^3 min^{-1}.
Sample: 18 ng of phosphate applied to column in each case as trimethylsilyl ester (after reaction of aqueous phosphoric acid or of ammonium phosphate with BSTFA/TMCS for 1 h at 65 °C).

3.4 Electrochemical Detectors

Electrochemical detectors working on coulometric principles are especially suitable for detection and determination of some inorganic gases (Table 50). This kind of detector is characterized not by a high sensitivity but rather by the possibility of high selectivity as compared with conventional GC detectors likewise suitable for the gases listed (particularly TCD, less so FID; FPD too for S-containing gases). They have been used with success in analysis of air.

Table 50

Coulometric detection in gas chromatography of inorganic gases.

Compound	Detection limit	Ref.
Hydrogen sulfide, sulfur dioxide	10 ppb, 50 ppb	[455]
Phosphorus hydride	5 ng absolute	[489]
Oxygen, nitrogen dioxide, carbon dioxide	ppm range	[599]
Oxygen	10^{-6} cm^3 absolute	[524]
Carbon monoxide	1.2×10^{-9} g	[756]

3.5 Radiometric Detectors

Measurements of radioactivity in flow-through cells for detection of tritium-labeled, fluorinated β-diketonates of metals give detection limits of around 10^{-12} g (= 1 pg). In one particular case, sulfur dioxide was allowed to react with labeled fluorine (^{18}F) to give $^{18}F_2SO_2$ and determined radiometrically after separation by gas chromatography with a detection limit of 10^{-18} g [467]. Compared with element-specific detectors, however, this detector plays a negligible role in gas chromatography of inorganic substances. The applications are compiled in Table 51.

Table 51

Radiometric detection in gas chromatography of inorganic substances.

Element: Compound	Detection limit	Ref.
SO_2 as $^{18}F_2SO_2$	10^{-18} g	[467]
Rare earths: ^3H-labeled hepta-fluoro-7,7-diethyl-4,6-octanedione chelates	1 – 4 ng	[477]
Al, Cr, lanthanides: ^3H-labeled polyfluorinated acetylacetonates	10^{-12} g	[506]

3.6 Comparison of Various Detectors

As long ago as 1964, *Albert* [461] was able to demonstrate that the detection limit for chromium trifluoroacetylacetonate with an ECD was two orders of magnitude lower than with a FID. In many cases, this ratio is nowadays more favorable for ECD by a further power of ten.

Berck and co-workers compared microcoulometric, thermionic, and flame photometric detectors as applied to **phosphine** [489]. They based their "detection limit" upon a 10% full scale deflection of the recorder at maximum sensitivity setting of the instrument. The flame photometric detector proved to be three orders of magnitude more sensitive than the microcoulometric detector.

Sowinski and *Suffet* demonstrated the high sensitivity of the ECD, as compared to microcoulometric and flame photometric detectors, for the case of boron hydrides. Gas volumes of 10 cm^3 were used in GC analysis.

Detection of various elements (in the form of organometallic compounds) has been performed with a microwave emission detector (MED) and with two FID

Table 51 a

Comparison of sensitivity of various detectors used in GC analysis of inorganic substances.

Element: Compound	Detection limit			Ref.
Cr: Trifluoroacetylacetonate	FID: 6×10^{-8} g ECD: 2.1×10^{-10} g			[461]
P: Phosphorus hydride	Microcoulom. D.: 5 ng TID: 20 pg FPD: 5 pg			[489]
B: Boron hydrides B_2H_6	Microcoulom. D.: 0.1 ECD: 0.01 FPD: 0.1 (data in ppm for 10 cm^3 sample)			[737]
Organometallic compounds	MED	FID*	FID**	
Si	0.8	0.4	–	
Fe	5.9	2	0.8	[800]
Pb	33	40	5	
Sn	0.9	5	5 [in ng]	

* air-rich, ** hydrogen-rich flame.

versions and the results compared. The sensitivity was found to vary according to the element; the selectivity of the MED surpasses that of the FID by 10^3 to 10^4.

The choice of a GC detector for determination of an element or an inorganic substance depends upon both the required sensitivity and the selectivity dictated by the particular analytical problem at hand after optimized gas chromatographic separation: it represents a compromise between maximum attainable sensitivity and maximum attainable selectivity, since both requirements are rarely met by a single detector.

III Applications to Inorganic Trace Analysis

The numerous possible modes of detection considered in Chapter II in conjunction with chromatographic separations of inorganic compounds have led to a series of powerful applications of chromatography in inorganic trace analysis.

The principal features of these procedures can be classified into the following four groups:

1) Application of a method of separation prior to actual quantitative analysis frequently **reduces** the **effort** to be invested **in sample preparation** (clean-up) for various materials. In the most favorable case, chromatographic separation constitutes the sample preparation step.

2) **Interference** to be expected on direct application of element-selective methods of determination in various materials can be **eliminated** on removal of interfering materials by chromatography.

3) The performance of many chromatographic methods permits **simultaneous determination** of many trace elements by separation in a single operation (multi- or oligo-element analyses).

4) Gas and liquid chromatographic methods provide a possibility of **enriching** traces of elements prior to detection and determination. Enrichment is always involved if the sample volume is greater than the elution volume (see Section III/3).

The followed detailed tabulations and the text concerning the applications of chromatographic methods to inorganic trace analysis with the techniques of paper, thin-layer, column, and gas chromatography should provide a survey of the versatility and performance of these methods; no claim is made to complete coverage.

Abbreviations used (in Tables 52 – 59):

A, alumina; E, elution; Em(is), emission spectrum analysis; F, fluorometry; IE, ion-exchange; S, silica gel; CE, cation exchange; Compl., complexometry; NA, neutron activation analysis; Ph, photometry; RD, reaction detector; Ref., reflectance measurements; RP, reversed phase chromatography; TFA, trifluoroacetylacetone; Tit., titrimetry; Trans., transmission measurements; P, partition chromatography; Vis., visual detection; Vol., volumetry; C, cellulose.

See Section II/3 for abbreviations of GC detectors.

1 Inorganic Materials

1.1 Paper Chromatography

Paper chromatography is the simplest of all chromatographic techniques with regard to execution of separations and it incurs the lowest costs for materials. The modest performance and rather poor reproducibility of separation have limited the number of practical applications. Problem-oriented procedures were published mainly before 1970, a striking number of these publications originating from East European and Russian (and also Chinese) sources.

Geochemical analysis of traces of elements is a preferred area of application in the field of inorganic materials. General information regarding time requirements, sensitivity, and reproducibility, also compared with other methods of element analysis, on

Table 52
Chromatographic analysis of inorganic materials – paper chromatography (selection).

Elements	Material	Method of determination	Detection limit	Ref.
Cu	Cadmium selenide	Vis (as carbamate)	0.05%	[225]
Nb	Ores, rocks	Vis (+ tannin)	50 ng	[96]
U	Ores, rocks	Vis: F	1 ng	[140]
U	Geochemical samples	Vis (+ PAN)	1 ppm	[320]
Al, Cu	Corrosion products	E – AAS	0.01 μg	[379]
Os, Ru	Mineral crude products	Vis (+ thiourea)	0.1 – 0.3 ppm	[334]
Se, Te	Minerals	in situ: Re (+ $SnCl_2$)	1 ppm	[128]
Se, Te	Sulfidic ores	E – Ph	μg range	[318]
Zr, Hf	Alloys	E – Ph (+ arsenazo III)	μg range	[97]
Bi, Cu, Pb	Synthetic mixtures	E – Ph	4 ppm	[19]
Co, Cu, Fe, Ni	Sulfidic rocks	Vis (+ rubeanic acid)	2 ppm	[434]
Co, Cu, Mn, Ni	Silicates	E – Ph	–	[88]
Co, Cu, Ni, Zn	Soil, rock samples	Vis, E – Ph, Em	5 – 20 ng	[266]
Na, K, Rb, Cs	Minerals	E – Ph	–	[22]

(Abbreviations, see p. 112)

use of paper chromatography in routine analysis will be found in the publication by *Arginier,* 1970 [3].

Semiquantitative evaluation of paper chromatograms can be performed after spraying with the aid of reference solutions; quantitative analyses are generally carried out after elution or after digestion of the paper (usually after prior location) by various methods of determination (see Table 52).

Cu: After selective extraction with dithizone the complexes are decomposed, Cu is separated from other elements and made visible as diethyldithiocarbamate. The absolute detection limit lies at 40 ng, and the reproducibility of visual evaluation for 1.4 μg is reported to be 0.07 μg (= 5%) [225].

Nb: Digestion with hydrofluoric and nitric acid; determinations performed in the range 2 – 5 μg [96].

U: Uranium can be determined fluorometrically after treatment of the paper with Na_3PO_4 and zinc acetate in the ratio of $1:10^6$ alongside other elements [140] or after spraying with the locating reagent PAN by comparison with standard solutions [320].

Cu, Al: A reproducibility of 5% can be attained by AAS for 0.1 μg Al or 1 μg Cu after elution [379].

Os, Ru: After digestion with $HClO_4/NaBiO_3$ the elements can be volatilized as tetraoxides, collected in ethanol/hydrochloric acid, and then separated by partition chromatography. Reaction with thiourea is performed at 70 °C [334].

Se, Te: After digestion of the various materials, multistep removal of interfering elements must be performed before paper chromatography. Separation is followed by reduction to the elements on the paper by spraying with a $SnCl_2$ solution and subsequent quantitative photometric analysis in the reflectance mode [128]. In another procedure, tellurium is eluted as $TeBr_4$ and selenium is eluted after reaction with thiourea and then determined photometrically in solution [318].

Hf, Zr: Paper impregnated with 10% ammonium nitrate is used as stationary phase for separation of these elements. Standard deviations of 0.5 and 1.9%, respectively, are reported for the determination of

17 µg of each of the elements. Zr and Hf can be separated in a ratio of 10:1 to 50:1 by this procedure (Nb interferes) [97].

Bi, Cu, Pb: The elements are separated on Whatman No. 1 paper with ethanol/5 N HCl (9:1) as mobile phase, located with 8-hydroxyquinoline/ammonia, and determined photometrically in solution at 270, 275, and 325 nm (Pb, Cu, Bi) after elution with conc. HCl [19].

Co, Cu, Fe, (Mn), Ni: Visual methods (rubeanic acid as spray reagent) [434] or photometric methods after elution [88] are employed for determination of these elements after separation by paper chromatography.

Co, Cu, Ni, Zn: Benzidine/rubeanic acid (for Co, Cu, Ni) and dithizone (for Zn) are employed. The standard deviations are 10%. Apart from colorimetry, emission spectral analysis can also be used [226].

Na, K, Rb, Cs: The alkali metals are located with violuric acid, the complexes eluted, and assayed by photometry at 550 nm [22].

In line with the present-day diminished significance of this chromatographic technique, only a few publications have been evaluated for this survey. They should suffice, however, to give an impression of sample materials, determination methods, detection limits, and reproducibility of quantitative analyses.

1.2 Thin-Layer Chromatography

As in paper chromatography, the removal of interfering substances prior to quantitative analysis plays a central role in the application of thin-layer chromatography to trace analysis of elements in inorganic materials.

Increased versatility in the separation systems by methods of adsorption, partition, reversed phase, and ion-exchange chromatography, as well as greater separation performance, also permit a series of multielement analyses. Table 53 surveys the potential applications.

Al: Thin-layer chromatography serves for removal of excess reagent from Al-8-hydroxyquinoline complex, and leads to an improvement of the detection limit. The Al-oxinate can subsequently be determined photometrically in solution at 260 nm [61].

Table 53
Chromatographic analysis of inorganic materials – thin-layer chromatography (selection).

Elements	Material	Chromatography/ Method of determination	Detection limit	Ref.
Al	Steel	Various M./E – Ph	ng range	[61]
In	Gallium	RP/F (+ rhodamine 6G)	0.18 ppm	[182]
Li	Alkali salts	C/F, Ph	1 ppm	[276]
Pb	Ceramic materials	C/Vis (+ diphenyl-carbazide)	20 μg	[8]
Te	"Cinnabar" (Hg-containing material)	C/E – Ph	10 ppm	[299]
U	Natural uranium rock	IE/Vis (+ PAN)	1 μg	[344]
Cu, Fe	Beryllium chloride chalcopyrite	C-RP/Planimetry	μg range	[169]
Ni, Co, Mn, Cr	Cosmic dust	S/Ref.	μg range	[338]
Fe, Cu, Hg, Zn, Ni, Ca	Rare earths	S/Vis	100 ppm	[425]
Rare earths	Uranyl nitrate	S/E – Ph	0.1 ppm	[435]
Rare earths	Magnetic alloys	S-RP/E – Tit, Ph	μg range	[364]
Rare earths	Rocks, techn. products	S-P/in situ	50 – 100 ng	[167, 201]
Rare earths	Uranium	C/Em	ng range	[199]
40 Cations, 15 anions	Alloys	S, A/Vis	0.06 – 6 μg	[153]

In: Indium is sorbed from 0.8 N HBr solution in RP system with tributyl phosphate (on "Ftoroplast" support) while the gallium matrix migrates with the front. A relative standard deviation of 7.5% has been reported for determination of In in 99.99% Ga. After elution of the In, the rhodamine 6G complex is extracted in benzene [182].

Li: Lithium is located on a cellulose layer by spraying with a uranyl zinc acetate solution and fluorescence measured directly on excitation at 366 nm [276].

Pb: The stationary phase for separation of lead consists of a mixture of carboxymethyl (CM) cellulose (= ion exchanger) and pure cellulose, and a buffer solution of pH is used as mobile phase [8].

Te: Tellurium can be separated from mercury with a mixture of benzene and tributyl phosphate as mobile phase; Hg migrates with the solvent front. Tellurium is located by spraying with tin(II) chloride in hydrochloric acid and determined photometrically after elution with bismuthiol II.

U: Ion-exchange chromatography on TLC plates is suitable for separation of uranium at contents of less 0.1%. Gallium can be detected simultaneously by this method [344].

Cu, Fe: The RP system consists of cellulose with thiothenoyltrifluoroacetone as stationary phase and an aqueous pyridine solution (pH 5) as mobile phase. Photometric determination performed at 390 (Cu) and 810 nm (Fe) [169].

Ni, Co, Mn, Cr: These metals can be separated with acetone/hydrochloric acid as mobile phase, and determined on the layer with standard deviations of 9 – 22% after spraying with rubeanic acid. Magnetic fractions of cosmic dusts have been investigated [338].

Fe, Cu, Hg, Zn, Ni, Ca: Rubeanic acid and 8-hydroxyquinoline have been used as locating reagents in TLC of these elements [425].

Rare earths: TLC of this group of elements has been examined particularly thoroughly. These studies led to systems permitting separation of all the rare earths. TLC separation is performed after extraction from 5 N HNO_3 with tributyl phosphate in benzene from uranyl nitrate solutions. Photometric determination is carried out with arsenazo III and M, respectively at 655 and 640 nm on the eluted samples [435].

The RP system consists of bis(2-ethylhexyl)phosphoric acid on silica gel and aqueous hydrochloric acid. The major components Cu and Co of magnetic alloys are removed with 0.1 N HCl as solvent, and the rare earths with 7 N HCl. Complexometric titrations and photometric methods are used for quantitative analysis [364].

Partition chromatography on silanized silica gel with diisopropyl ether, tetrahydrofuran, and 65% HNO_3 (100:70:6, v/v) permits individual determination of rare earths in various materials. Spraying with aqueous neothorin solution and exposure to ammonia serves for visual detection and determination [167, 201].

After removal of uranium, emission spectral analysis with photographic evaluation gives detection limits of 1 – 100 ng with relative standard deviations between 3 and 30% [199].

A **solvent extraction system** with subsequent TLC separation has been proposed for individual determination of numerous **cations and anions** in solutions of alloys [153]. The elements in the various groups of the system can be readily separated from one another and can be determined with a wide variety of reagents. The publication cited gives a systematic and very detailed account of this TLC analysis.

1.3 Column Chromatography

The principal significance of column chromatography without direct detection in a flow-through detector (as off-line procedure) lies in the enrichment, and the pre-separation usually associated therewith, of traces of elements for subsequent quantitative analysis.

Apart from ion-exchange chromatography, various partition or reversed phase chromatographic methods have been applied to these tasks. Methods suitable for determination of analytes after such chromatographic sample preparation include AAS, voltammetry, neutron activation analysis, and photometric or fluorometric and also volumetric (complexometric) procedures (see Table 54a). Fractionation of the eluate may, in special cases, permit separation of several elements, making less selective methods also suitable for quantitative analysis.

The development of column-chromatographic techniques in direct connection with a flow-through detector (on-line procedures) has considerably simplified and shortened analysis. In spite of the greater separation performance of HPLC relative to traditional column chromatography, only few procedures have so far been described for determination of several elements in a single analytical step (see Table 54b). Spectroscopic and electrochemical detectors have proved best for detection in flow systems.

Table 54a

Chromatographic analysis of inorganic materials – column chromatography (off-line procedures).

Elements	Material	Chromatography/ Method of determination	Detection limit	Ref.
Be	Geological, industrial materials	IE/AAS	ppb range	[228]
Bi	Cerussite, pure Pb	RP/AAS and IE/AAS	ppm range	[381, 382]
In	Gallium	RP/F	18 ppm	[182]
Sn	Sn-Cu mixtures	C/Vol	μg range	[99]
Sn	Rocks, soil, ferromanganese materials	IE/Ph	0.1 ppm	[369]
Tl	Silicates	IE/F, NA	0.1 ppm	[282]
U	Irradiated products	RP/NA	0.5 ng absolute	[437]
Al, V	Titanium	IE/NA	0.15 ppm	[301]
Bi, Pb, Zn	Fe-alloys	IE/Ph	2 ppm	[235]
Cd, Cu, Pb	Pure bismuth	CE/Voltammetry	10 ppm	[366]
Cd, Hg, Zn	Copper	RP/Ph	ng range	[262]
Lu, Yb, Tb	Rocks	IE/NA	ppb range	[56]
Pb, Th, U	Tantaloniobate	IE/Compl, Ph	ppm range	[400]
Cr, Cu, Fe, Pb	Synthetic mixtures	IE/Vol	10^{-6} mol/dm^3	[312]
Ca, Mg, Na, K	Tungsten	IE/AAS	1 – 20 ppm	[231]
Cd, Cu, Co, Pb Zn, U	Manganese nodules	IE/AAS	ppm range	[229]
Pt-elements	Iron meteorites	IE/Ph	ppb range	[346]
Alkalis, alkaline earths	Molybdenum	Sephadex/Emis, AAS	ppm range	[205]
Rare earths	Uranium	C/Emis	ng range	[199]
Numerous metals	Enriched uranium solutions	RP/AAS	ng range	[313]
15 Elements	Silicates	IE/AAS	ppm range	[284]
22 Elements	Silicates	IE/Vol, Ph, AAS	ppm range	[401]
Numerous elements, e.g. Ir, Ba	TiO_2 crystals	IE/NA	Ir: 1 pg, Ba: 100 ng	[302]

Table 54b
(On-line procedures including HPLC methods.)

Elements	Material	Chromatography/ Method of determination	Detection limit	Ref.
Bi	NBS materials	IE/UV-254 nm	10 ppb	[441]
Fe	NBS materials	IE/Ph-355 nm	10^{-2}%	[354]
Pb	Standards	IE/UV-270 nm	ng range	[353]
Sb	Alloys	IE/Coulometry	1 ppm	[409]
Re	Molybdate concentrates	RP/UV-227 nm	20 µg absolute	[308]
Sn	NBS materials	IE/UV-225 nm	100 ppm	[114]
U	Rocks	A/RD-Ph	µg range	[380]
Alkalis, alkaline earths	Synthetic mixtures Zr phosphate	IE/FID	10^{-8}g (Cs) 10^{-10} – 10^{-12}g	[11] [10]
Ni, Pb, Zn	NBS standards	IE/RD-Ph	1% and below, µg range	[113]

Be: After extraction with acetylacetone, separation is performed on an acid cation exchanger by adsorption from the tetrahydrofuran/chloroform/methanol/hydrochloric acid system. Elution is then possible with 6 M HCl. Recovery of 0.063 – 0.63 µg of Be is 95 – 100% [228].

Bi: Bismuth is separated from an excess of Fe, Sb, and Cu as thiourea complex on polymer + tributyl phosphate [381].

The element can be separated from lead by ion exchange between EDTA-containing 0.3 – 0.7 M perchloric acid and hydroxylammonium chloride-containing solution, and determined at Ge contents of 2×10^{-4}% in lead and of 2×10^{-3}% in galenite [382].

In: Separation of the gallium matrix (99.99%) is performed on "Ftoroplast" polymer with tributyl phosphate. Traces of indium are determined after elution by fluorometry with rhodamine 6G from 0.8 N HBr after benzene extraction. A relative standard deviation of 7.5% is reported for 1.8×10^{-5}% of In [182].

Sn: Tin can be separated from copper by paper and column chromatography in the cellulose/butanol-HCl system. Volumetric determination is carried out with potassium iodate in the eluate [99].

Tl: Anion-exchange chromatography serves for preconcentration and selective separation. Variation coefficients of 1.2% can be achieved on fluorometric analysis with rhodamine in the 6 ppm range [282].

U: Traces of uranium can be separated from hydrochloric acid solution on Porapak with 2-thenoyltrifluoroacetylacetone as chelating agent. Recovery in the nanogram range is 96.8 ± 1.6% [437].

Al, V: Separation of titanium by anion-exchange chromatography has been compared with precipitation as hydroxide [301].

Bi, Pb, Zn: Separations by anion-exchange chromatography are followed by photometric determinations with dithizone or as iodo complex; standard deviations in the ppm region are 5 – 20% [235].

Cd, Cu, Pb: The ethylenediamine complexes of these elements are separated on an anion exchanger; bismuth passes through the column without retention after addition of tiron. Voltammetric analysis after fractional elution show standard deviations less than 15% [366].

Cd, Hg, Zn: After treatment with dithizone in o-dichlorobenzene, a Chromosorb material can be used for separation of heavy metals by displacement chromatography [262].

Lu, Yb, Tb: A mixed solvent system has been used for enrichment and fractionation in anion chromatography of these rare earths [56].

Pb, Th, U – Cr, Cu, Fe, Pb: Volumetric (complexometric) and photometric methods have been employed after fractional ion-exchange chromatography for Pb, Cr, Cu, Fe and for Th (thoron) and U (PAN) [400, 312].

Ca, Mg, Na, K: Matrix removal (tungsten) performed on DEAE Sephadex [231].

Cd, Cu, Co, Pb, Zn, U: These elements are adsorbed from concentrated hydrochloric acid on a strongly basic anion exchanger in admixture with isopropanol and hexane, and can be eluted stepwise with acids of various concentrations [229].

Pt elements: Photometric determinations are performed with various reagents after separation of iron [346].

Alkali and alkaline earths metals: Molybdenum matrix can be adsorbed onto Sephadex as Mo-NH_3-glucose complex [205]. Alkaline earth metals determined by AAS.

Rare earths: Uranium can be removed by a combination of partition chromatography on cellulose and diethyl ether/nitric acid partitioning (extraction). Standard deviations of 3 – 30% can be attained in the range of 1 – 100 ng of the rare earths [199].

Numerous metals: Uranium can be separated on Kel-F charged with tributyl phosphate [313].

15 elements: Stepwise elution from a cation exchanger leads to separation of major and minor components and traces in silicate rocks for subsequent AAS analysis [284].

22 elements: Cation and anion exchangers used for separation by stepwise elution [401].

Numerous elements, e. g. Ir, Ba: Combination of anion exchange and extraction with tributyl phosphate leads to isolation of 22 contaminants in titanium dioxide crystals [302].

On-line procedures (detection in flow system).

Bi: Bismuth can be isolated from other metals such as Hg, Cd, Sb on cation exchangers such as Dowex 50W-X8 or Amberlyst 15 by elution with 0.5 M HBr, and determined directly by UV absorption measurements in the effluent [441].

Fe: An anion exchanger (Amberlyst A 26) is used for separation of Fe(III) as chloro complex from 6 M HCl. Detection of this complex is possible in a flow-through photometer. The relative standard deviation is 4.1% for contents of 0.07% [354].

Pb: Like Fe (vide supra), lead can be separated from 8 M HCl on an anion exchanger and determined by photometry in a flow system. Duration of this analytical procedure (also for Bi, Fe – vide supra), which has been called forced-flow liquid chromatography, is only 8 min [353]. Enrichment and separation are coupled together.

Sb: A coulometric detector has been used for determining antimony by the technique given above for Bi, Fe, Pb. The principle of detection is an irreversible electrochemical oxidation of Sb(III) on a Pt electrode [409].

Re: ReO_4^- and MoO_4^{2-} ions are initially adsorbed to varying degrees on teflon with tributyl phosphate, and thus separated. Perrhenate can be eluted with water [308].

Sn: Selective separations of cations can be accomplished on macroporous polyacrylate exchangers with an amide group (XAD-11). Tin is eluted with 0.1 M HCl and, after admixture of conc. HCl, determined automatically by photometry in a flow-through cell as chloro complex in 10 M HCl. Relative standard deviations of 1% are found for materials with Sn contents of 1% [114].

U: Uranium is adsorbed selectively onto silica gel from EDTA and tartaric acid solutions of pH 5, and is subsequently eluted with 3 M acetic acid. After addition of arsenazo III the uranium can be determined by photometry of the eluate in a flow-through cell [380].

Alkali, alkaline earth metals: Ion-exchange chromatography on zirconium phosphate and detection in a special flame ionization detector are coupled directly via a sample collector for the eluate [11, 10].

Ni, Pb, Zn: Use of various metal reagents such as arsenazo I, III or PAR in a reaction detector permits photometric determination of cations separated by ion-exchange chromatography [113].

1.4 Gas Chromatography

Owing to the numerous possibilities of complex formation of metals (see Section I/2.2.3) with selectivity effects prior to gas chromatographic separation, and owing to the many very sensitive detectors available for gas chromatography, this method has also found wide application in trace analysis of elements in inorganic materials. Moreover, a series of volatile inorganic substances can be separated and determined by gas chromatography.

A selection of publications presented in Table 55 will illustrate these possibilities.

Table 55 Chromatographic analysis of inorganic materials – gas chromatography.

Elements	Material	Compound/ Detector	Detection limit	Ref.
As	Oxides, sulfides, rocks, alloys	$AsCl_3$/TCD	ppm range	[596]
Be	Lunar rocks, meteorites	$Be(TFA)_2$/ECD	40 ppb	[540, 734]
Be	Alloys	$Be(TFA)_2$/MED	1 pg absolute	[706]
Cl	$SiBr_4$	Cl_2, $SiCl_4$/MS, ECD	300 ppm HCl, 0.2 ppm $SiCl_4$	[458]
Cr	Iron alloys	$Cr(TFA)_3$/ECD	20 ppb	[690]
F	Water, phosphates	$(C_2H_5)_3SiF$/FID	ppb range	[495]
Ge	Oxide, ores, alloys	$GeCl_4$/TCD	10 μg absolute	[597]
P	Text mixtures	PH_3/TID	–	[573]
Sb	$GaCl_3$	$SbCl_3$/AAS (flame, flameless)	50 and 3 ppm	[695]
Se	Steel	5-Chloropiaz-selenole/TCD	0.1%	[459]
Se	Elemental As, As_2O_3	5-Nitropiaz-selenole/ECD	3 ppm	[460]
Se	Pure sulfuric acid	5-Nitropiaz-selenole/ECD	ppm range	[727]
Se	Highly pure copper	5-Nitropiaz-selenole/ECD	1 ppb	[646]
Si	Alloys	$SiCl_4$/TCD	ppm range	[660]
Si	Iron, steel	$SiCl_4$/TCD	50 ppm	[729]
Si	Nonferrous alloys	$SiCl_4$/TCD	less than 1%	[730]
Sn	Nonferrous alloys	$SnCl_4$/TCD	less than 1%	[730]
Sn	Zircaloy	$SnCl_4$/TCD	ppm range	[476]
Sn	Oxides, sulfides, minerals	$SnCl_4$/TCD	μg range	[598]
Al, Cr	Uranium	TFA derivative/ ECD	0.1 ppm	[561]
Cu, Al	Zinc	TFA derivative/ MED	20 ppb	[704]
NO_3^-, NO_2^-	Aqueous test solutions	Nitrobenzene/ECD	0.12 ppm	[563]
Rare earths	Extraterrestrial material	Heptafluoroke-tonates/radiom. D	0.04 ppm	[477]
CO_2	Rocks, minerals	CO_2/TCD	5 ppm	[604]
CO_2, CO, COS, H_2S	Helium	Direct/He-detector	0.01 – 5 ppm	[503]
He, H_2, Ar, O_2, Kr, N_2, Xe	Pure helium	Direct, as gases/ TCD	ppb range	[675]

As: Arsenic-containing material is chlorinated with carbon tetrachloride in a sealed glass tube, and the volatile products are transferred, after breaking of the microvessel in a special device, directly to the gas chromatograph [596].

Be: Standard samples containing 0.63 ppm of beryllium were determined with a standard deviation of 0.05 ppm. Comparison with a fluorometric method of determination gave average values in good agreement; however, a lower standard deviation was found in the latter case [540]. The absolute detection limit is reported as 4×10^{-14} g [734].

Be: Samples of alloy are dissolved in hydrochloric acid and extracted with HTFA (trifluoroacetylacetone) at pH 6.5 after addition of EDTA to mask aluminum [706].

Cl: Impurities in $SiBr_4$ are identified by mass spectrometry in combination with gas chromatography, determination of chlorine in the presence of methane, which were not separated by gas chromatography can be accomplished by simultaneous FID/ECD detection [458].

Cr: Sample treated directly with reagent after addition of catalytic amounts of HNO_3 and heating. The chromium complex is then extracted from aqueous solution with benzene [690].

F: Fluoride ions can be reacted with variously substituted silanols in strongly acid solution, and extracted, for example, as trimethylfluorosilane with organic solvents. Quantitative analysis by GC gave detection limits of 200 ng for a TCD and 0.6 ng for a FID. The standard deviation was 6% for 0.95 $\mu g/cm^3$. Interferences come only from Zr, Th, V, and high concentrations of silicic acid [495].

Ge: Sample material is treated as for arsenic (see arsenic) [597].

P: Substances containing phosphorus, such as tributyl phosphate, are decomposed to phosphorus hydride on quartz at 900 – 1000 °C [573].

Sb: Coupling of GC and AAS permits selective and sensitive determination of antimony (at 217.6 nm) in the presence of gallium. Atomization in a graphite tube cell (flameless AAS) is performed at 2000°C [695].

Se: Numerous selenium determinations by gas chromatography after reaction of aqueous solutions with substituted aromatic diamines to give 5-chloro- and 5-nitropiazselenole permit sensitive determination of this element in a very wide range of materials [459, 460, 646, 727].

Steel samples are treated with a mixture of $HCl/HNO_3/HClO_4$; Fe(III) is masked with phosphoric acid before the reaction [459]. For sample weights of 0.1 g with contents of 0.1 – 0.3%, standard deviations of 0.002 – 0.003% can be attained with a TCD [459].

Arsenic materials (1 g sample) are treated with HNO_3/H_3PO_4 mixtures (20:5, v/v) and the resulting 5-nitropiazselenole is extracted with toluene. A recovery of 99.9% and a standard deviation of 0.18 μg were found for 3.76 μg/g [460].

Selenium contents of 0.15 $\mu g/cm^3$ were found in concentrated **sulfuric acid** [727].

A special procedures has been described for determination in **ultrapure copper**, which involves initial volatilization in a current of oxygen [646].

Si: Samples treated, as for arsenic (see arsenic), with carbon tetrachloride [660]. Determinations in nonferrous alloys gave standard deviations of 0.01 to 0.03% (absolute) for contents of 0.7% [730].

Sn: Results of GC determinations of tin in zircaloy have been compared with those of X-ray fluorometric analysis and polarography: comparable mean values were obtained, with the GC technique showing the lowest standard deviation [476].

Al, Cr: GC analysis of these elements gave an absolute detection limit of 1 pg; not only GC but also photometric determinations have been performed in the ppm region [561] (see also Section III/5).

Cu, Al: The trace contents of copper and aluminum in a 5 g sample of zinc have been extracted as chelates from aqueous solution at pH 4.5 with chloroform (recovery 95%). The absolute detection limits are 1 ng for Cu and 0.5 ng for Al [704].

NO_3^-, NO_2^-: Nitrite is oxidized to nitrate by $KMnO_4$. Reaction with benzene to give nitrobenzene is performed in a 3:1 mixture of sulfuric acid and water. The recoveries are 90 ± 8% and 92 ± 12%, respectively, for 0.12 – 62 ppm of nitrate and 0.09 – 37 ppm of nitrite [563].

Rare earths: The elements Yb, Ho, Dy, Tb, Lu are separated as tritium-labeled fluorinated ketonates and determined with a radiometric detector in absolute quantities of 1 – 4 ng [477].

Inorganic gases, either elements or compounds mainly of carbon, nitrogen, or sulfur, are determined in a very wide range of inorganic materials:

In a solid matrix: iron and steel, aluminum, beryllium, irradiated BeO, high-melting metals, metals in general, metal surfaces, thin layers, iron, oxidic materials minerals, glass (according to [701]).

In a liquid matrix: water.

In a gaseous matrix: nitrogen, nitrogen oxides, inert gases, carbon dioxide, carbonation gas, etc. (see [701]).

The following gases warrant special mention: CO, CO_2, N_2O, NO, NO_2, N_2, H_2S, SO_2, COS, H_2, Ar, O_2, Cl_2, He, Ne, HCl.

CO_2: Carbon dioxide is liberated from carbonate-containing materials by treatment with phosphoric acid; sample weights of 1 g are used [604].

CO_2, CO, COS, H_2S: Use of a helium ionization detector (with tritium foil) permitted detection of 0.01 ppm of CO_2, 1 ppm CO, 2 ppm COS, and 5 ppm of H_2S in 0.2 cm^3 of helium after separation by GC [503].

Ne, H_2, Ar, O_2, Kr, N_2, Xe: Enrichment of these gases by various methods likewise permits attainment of detection limits below 1 ppm with a TCD [675].

Further applications of gas chromatography to inorganic analysis (up to 1971) will be found in the review by *Rüssel* and *Tölg* [701].

2 Organic, Biological, and Special Environmental Materials

Chromatographic analysis of elements or inorganic materials in organic, biological, or special environmental materials generally makes greater demands upon sample preparation than in inorganic materials.

Nevertheless, for a series of analytical problems and for determination of a number of elements, chromatographic methods remain superior to element-selective ones requiring less sample preparation. The chemical derivatization of the elements or inorganic substances prior to chromatographic separation constitutes an important step in sample preparation and eliminates many sources of interference.

Furthermore, chromatographic methods provide a means of separating various compounds of an element in a given material and determining them separately, with detection in direct combination with the column playing a very important role (see also Section III/4: Elemental Species).

The following **Tables 56 – 59** survey possible applications to a wide variety of materials. Demonstration of versatility rather than exhaustive citation has been the guiding principle adopted in selecting material for these compilations.

2.1 Paper Chromatography

Methods for the determination of traces of elements in complex organic matrices are generally made up of several individual steps. In such a system paper chromatography has a particular task to fulfil, i.e. removal of ions which would interfere with subsequent quantitative analysis.

Most separations by paper chromatography are preceded by separation and enrichment of the elements from the sample material by extraction of metal chelates; partition chromatography then requires transformation into ions by decomposition of the metal complexes.

Analytical procedures can be considerably simplified if direct in situ evaluation of the paper chromatograms in the reflectance mode is possible. Otherwise a fresh digestion step, that of the paper with the separated elements (after location), will be necessary; metals or compounds can rarely be eluted readily from the cellulose.

Table 56 gives a selection of such paper chromatography methods.

Table 56

Chromatographic analysis of organic, biological, and special environmental materials – paper chromatography (selection).

Elements	Material	Method of determination	Detection limit	Ref.
Cu	Pharmaceuticals	E – Ph	4 μg/sample	[371]
Fe	Plants, organic materials	Vis	7 – 35 μg absolute	[127]
Pb	Vegetables	Extraction – Ph	μg amounts	[35]
Sn	Foodstuffs	E – Ph	0.4 – 1.5 μg/ 15 cm^3	[2]
Sr	Vegetables, seawater	Extraction – Ph	μg amounts	[34]
Tl	Organic material	Extraction – Ph	μg amounts	[37]
U	Plants, soil samples	Ph, F	2 ppm	[325]
U	Natural waters	Vis – F	1 ng absolute	[140]
V	Biological material	Extraction – Ph	μg range	[36]
Zn	Blood serum	Ph	μg/cm^3 serum	[66]
Zn	Blood	Extraction – Ph	2.5 μg absolute	[370]
Zn	Urine	E – Complexometry	3 μg absolute	[337]
Cu, Ni	Natural waters	Planimetry	2.5 mg/cm^3 Cu, 0.05 mg/cm^3 Ni	[39]
Co, Cu, Ni	Water	E – Ph	–	[73]
Cu, Fe, Mn, Zn	Plant leaves	Reflectance	1 μg absolute ppm range	[436]
Li, K, Rb, Cs	Mineral water	E – Flame photometry	–	[78]
Co, Cu, Mn, Mo, Ni	Foodstuffs	Reflectance	50 ng absolute, ppb range	[75]
Co, Cu, Fe, Mn, Mo, Ni, V, Zn	Vegetable oils	Ph	1 μg absolute	[136]
Traces of elements	Biological material	(Survey)	–	[101]

Cu: At contents exceeding 4 µg/sample, copper can be eluted from the paper as dithizonate and determined photometrically with a standard deviation of 1.3% [371].

Fe: In order to separate iron, paper is successively impregnated with $K_4Fe(CN)_6$ and $CoCl_2$. Determinations by weighing of paper spots after chromatography is possible in the range of 7 – 35 ng of Fe [127].

Pb: Lead is initially extracted as dithizonate and then separated on paper after decomposition of the complex. After ashing of the cut-out spots, photometric determination is performed with PAR [35].

Sn: Paper chromatography has the task of separating organic tin compounds from inorganic tin. There follows a column chromatographic separation of Sn-catechol complex from catechol violet reagent [2].

Sr: Strontium is located on paper with Na rhodizonate, then ashed with the paper, and determined photometrically with o-cresol-phthalein at 578 nm [34].

Tl: Thallium is first isolated from organic material by extraction with dithizone, and then separated from other elements on paper. Spraying with dithizone is followed by wet digestion of the paper with $HNO_3/HClO_4$ and extractive-photometric determination of Tl(III) with methyl violet [37].

U: A mixture of citric acid, acetone, ethanol, and HNO_3 is used for paper chromatography [325]. Detection is by fluoroescence after treatment of the paper with Na_3PO_4 and zinc acetate [140].

V: The procedure consists of four steps:
1) Extraction of vanadium with cupferron;
2) paper chromatography;
3) location with tannin and extraction from paper spot;
4) photometry with PAR [36].

Zn: Various possibilities have been described for the detection and determination of zinc on paper chromatograms:
a) Spraying with 8-hydroxyquinoline [66];
b) emission spectrometry after ashing [370];
c) localization with 8-mercaptoquinoline, elution and complexometric titration with sodium arsenate

and Eriochrome Black. The standard deviation lies below 10% for 30 – 90 μg of Zn [337].

Cu, Ni: The paper is impregnated with diacetyldioxime (for Ni) or dithioxamide (for Cu); evaluation via spot size [39].

Co, Cu, Ni: These elements can be located on paper by spraying with rubeanic acid and evaluated semi-quantitatively. Quantitative determination is performed after elution by photometry with nitroso-R-salt, dithizone, and diacetyldioxime, respectively, in solution [73].

Cu, Fe, Mn, Zn: Prior to paper chromatography, traces of these elements are separated and enriched in an ion-exchange column. Determination is on the paper for amounts below 2 μg, and by elution and photometry with various reagent in solution for amounts between 5 and 40 μg [436].

Li, K, Rb, Cs: The ions are separated by circular paper chromatography, eluted, and determined by flame photometry [78].

Co, Cu, Mn, Mo, Ni: After ashing of the sample material, the metals are separated as chlorides with various mobile phases and identified by colorimetry. Evaluation is performed with calibration solutions [75].

Co, Cu, Fe, Mn, Mo, Ni, V, Zn: Enrichment and partition chromatography on paper are followed by location of the elements with various reagents, elution, and determination with the aid of a capillary photometer [136].

Traces of elements: Pre-extraction and enrichment with 8-hydroxyquinoline is combined with separation by chromatography after decomposition of the complexes [101].

2.2 Thin-Layer Chromatography

In trace analysis of toxic elements by thin-layer chromatography, combination with extraction as **metal dithizonates** has proved particularly valuable.

Table 57 shows some relevant examples: silver [277], cadmium [279], mercury [12], lead [27], cadmium, cobalt, mercury, nickel, zinc [90], bismuth, copper, mercury, tin, zinc [446], silver, cadmium, cobalt, copper, nickel, lead, zinc [20], 14 metals – including oxinates [447].

With these metal chelates, thin-layer chromatography can be used for determination of a single element by removal of interfering compounds, or for multielement analysis in a single step.

Photometry in the reflectance or transmission mode is suitable for in situ evaluation of thin-layer chromatograms [277, 279, 446, 447]; it is also used after elution of the chelates (e.g. [12, 27, 90]).

The recovery of lead from blood is, for example, 99.1% for contents of 20 µg/100 cm^3 [27]. The standard deviations after elution lie between 7.8 and 12% [90].

Photometric determinations are performed at wavelengths between 400 nm for Ag [277] and 538 for Zn [90].

B: The boron-curcumin complex is separated from excess reagent before photometric determination [57].

Mg: Magnesium is separated by partition chromatography on alumina and located by spraying with benzidine [355].

Mn: The manganese-PAN complex is suitable for adsorption chromatographic separation of other interfering compounds or elements and for photometric determination in the reflectance mode at 576 nm [278].

Pb: Lead can be determined on cellulose TLC layers by fluorometry of the $PbCl_4^{2-}$ complex [420, 418] or visually after spraying with disodium tetrahydroxyquinone [147].

S: For determination of free sulfur in shampoo, sample preparation is by addition of 2,2′-dimethoxypropane and hydrochloric acid to the aqueous emulsion and extraction with carbon disulfide. Quantitative evaluation of the TLC plates is performed at 295 nm with a dual beam densitometer. The standard deviations lie below 5% in the range of 0.3 – 3 µg [315].

Alkaline earths: Fluorometric analysis is performed after spraying with 8-hydroxyquinoline and exposure to ammonia [280].

Table 57

Chromatographic analysis of organic, biological, and special environmental materials – thin-layer chromatography (selection)*.

Elements	Material	Chromatography/ Method of determination	Detection limit	Ref.
Ag	Drugs	S/Trans	50 ppm	[277]
B	Caviar	S/Vis	1 ng absolute	[57]
Cd	Pharmaceuticals	S/Refl	μg range	[279]
Hg	Foodstuffs	S/E – Ph	μg range	[12]
Mg	Cotton, cottonseed oil	A/Vis	0.1 – 0.3 ppm	[355]
Mn	Pharmaceuticals	S/Trans	0.5 ppm	[278]
Pb	Mineral oils	C/F	2 ppb	[420]
Pb	Lubricating oils	C/F	0.3 μg absolute	[418]
Pb	Air	Vis	μg amounts	[147]
Pb	Blood	S/E – Ph	20 μg/100 dm^3	[27]
S	Shampoo	S/Trans	0.3 μg	[315]
Alkaline earths	Vegetable (ash)	C/F	50 ng Ba, 20 ng Sr, 10 ng Ca, 5 ng Mg	[280]
Al, Be, Cr	Water	S/F, Ph	0.1 ppm Al, 1 ppm Be, Cr	[445]
Co, Cu, Ni	Grain	S/Trans	ppb range	[106]
Co, Cu, Ni, Mn	Pharmaceuticals	S/PAN complexes, E – Ph	10 – 35 μg/ 25 cm^3	[100]
Co, Cu, Fe, Ni	Water supply	A/Vis	5 ng Co, 8 ng Cu, Fe, 10 ng Ni	[110]
Cd, Co, Hg, Ni, Zn	Biological material, air	S/Dithizonates, E – Ph	1 μg absolute	[90]
Bi, Cu, Hg, Sn, Zn	Water	S/Vis, Trans	0.5 ppm	[446]
Co, Cu, Fe, Ni, Zn	Blood, tissue	S/Vis	μg range	[154]
Ag, Cd, Co, Cu, Hg, Ni, Pb, Zn	Blood, urine, excretia	S/Dithizonates – Ph	100 ng/dm^3	[20]
14 Metals	Water	S/Trans	0.1 ppm: Fe, Cu, Hg, Cd 0.5 ppm: Co, Ni, Mn, Pb, Zn, Bi, Sn 1 ppm: Al, Be 8 ppm: Cr	[447]

* See also: "Selenium in Water. An Example of Sub-ppb Quantitation in HPTLC" by *W. Funk* in *W. Bertsch, S. Hara, R. E. Kaiser*, and *A. Zlatkis*, "Instrumental HPTLC" (from the series "Chromatographic Methods"), Dr. A. Hüthig Verlag, Heidelberg-New York (1980).

Al, Be, Cr: In this procedure, aluminum and beryllium are likewise determined fluorometrically as the oxinates, and chromium after spraying with diphenylcarbamide [445]. TLC is preceded by extraction of the oxinates with subsequent re-extraction of the elements prior to separation by partition chromatography.

Co, Cu, Ni: Individual determination of these elements is performed with the aid of the spray reagents rubeanic acid or pyridin-2-aldehyde 2-quinolylhydrazone [106]. Visual detection limits for the first reagent lie at 30 ng (for Cu, Ni) and 50 ng (for Co), and for the second one at 10 ng per element and spot. A standard deviation of 20% is obtained for a Co content of 0.02 ppm, and 6.4% for a Ni content of 0.98 ppm [106].

Co, Cu, Ni, Mn: Partition chromatography of the PAN complexes after their extraction from aqueous solution permit individual photometric determination after elution [100].

Co, Cu, Fe, Ni: This procedure combines separation of pyridin-2-aldehyde 2-quinolylhydrazone derivatives by circular chromatography with the ring oven technique, leading to a very sensitive and selective trace analysis method for these elements [110].

Co, Cu, Fe, Ni, Zn: After dry ashing of biological materials at 700 °C interfering elements are removed by solvent extraction. Thin-layer chromatography is carried out with a mobile phase containing acetylacetone as chelating agent. Various spray reagents are used for detection [154].

2.3 Column Chromatography

Liquid chromatography in columns can be used for elemental analysis in biological as in inorganic materials for sample preparation (enrichment of traces and removal of interfering elements and compounds) with separate determination in the fractionated eluate or, to an increasing extent, for direct measurement in flow-through detectors.

Some examples of so-called off-line procedures are shown in Table 58a.

Ag: A combination of anion exchange and extraction with ammonium pyrrolidinedithiocarbamate permits enrichment of silver and its photometric determination [70].

Table 58a

Chromatographic analysis of organic, biological, and special environmental materials – column chromatography. (Off-line procedures, examples.)

Elements	Material	Chromatography/ Method of determination	Detection limit	Ref.
Ag	Water	IE/AAS	0.1 ppb	[70]
B	Water, nutrient solns.	IE/Ph	µg range	[264]
Ba	Water	IE/AAS	10 ppm	[365]
Ce	Seawater	RP/F	1 ng/dm^3	[359]
Sn	Foodstuffs	A (asbestos/C)/Ph	80 ppm	[2]
Tl	Marine sediments, seawater	IE/F, NA	0.1 ppm	[282]
Cd, Pb	Urine, blood	IE/AAS	1 ppm	[431]
Mo, W	Seawater	IE/Ph	ppb range	[207]
Cd, Co, Cu, Mn, U, Zn	Natural waters	IE/AAS, F	ppb range	[227]

B: Photometric determination is performed on the curcumin complex; interfering anions and cations are removed by ion-exchange chromatography [264].

Ba: Barium is enriched from water on a Dowex ion exchanger for determination by AAS [365].

Ce: Poly(vinyl chloride) with bis-(2-ethylhexyl) phosphate serves as stationary phase for enrichment of cerium. Sensitive fluorometric determination as cerium(III) ions at 255/350 nm is possible in the eluate after treatment with titanium (III) ions [359].

Sn: Column chromatography is used here to separate the Sn-catechol complex to be determined by photometry from excess catechol violet reagent [2].

Tl: Pre-concentration and almost specific separation of the element can be attained with aid of an anion exchange scheme. Determinations of ppb contents in water are performed by neutron activation analysis. Fluorometric determination with rhodamine is suitable for contents of 0.1 – 6 ppm (standard deviation for 6 ppm: 1.2%) [282].

Cd, Pb: These elements are enriched on an ion exchanger with a quaternary ammonium group and

Table 58b
On-line procedures, including HPLC*.

Elements	Material	Chromatography/ Method of determination	Detection limit	Ref.
Cr	Liquified coal	S/AAS	4 $\mu g/cm^3$ (mobile phase)	[195]
Cu	Water	IE/UV-235 nm, AAS	ng range	[196, 197]
Sb	Hair	IE/Coulometry	1 ppm	[409]
Pt	Urine	IE/UV-346 (RD)	1 ng absolute	[41]
S	Water, oil	RP/UV-254 nm	1 ng absolute	[64]
Se	Water	RP/UV-320 nm	0.32 ppb	[399]
Se, NO_2^-	Water, biological material	RP/UV, F	0.5 ppm Se, 0.1 ppm NO_2^-	[440]
NO_2^-, NO_3^-	Cereal stalks	IE/Coulometry	0.1 ppm (N)	[81]
Zn	Natural waters	IE/RD – Ph	ppb range	[281]
Monofluorophosphate	Toothpaste	IE/RD – Ph	0.1%	[24]
Cl^-, Br^-, I^-	Seawater	Ion chromatography/Potentiometry	ppm range	[104]
$S_2O_3^{2-}$, polythionates	Water	IE/RD – F	0.3 ppm	[442]
Anions and cations	Waste water, blood, urine, fruit juices	Ion chromatography/Conductometry	ppm range	[367]

* Concerning HPLC see also: *R. E. Kaiser* and *E. Oelrich*, "Optimization in HPLC", (in the series "Chromatographic Methods"). Dr. A. Hüthig Verlag, Heidelberg-New York (1981).

separated for AAS analysis. Recovery is 95% on average, and the relative standard deviations are about 10% [431].

Mo, W: The elements are adsorbed from 0.1 N HCl and 0.1 M NH_4SCN solution on a cation exchanger. Tungsten is separated from molybdenum on a DEAE cellulose anion exchanger. Photometric determinations with dithiol are possible in the ppb to ppt range [207].

Cd, Co, Cu, M, U, Zn: These elements are first extracted as dithiocarbamates. They are then adsorbed from a solution in tetrahydrofuran, methyl glycol, and 6 M HCl onto an anion exchanger, and subsequently eluted stepwise with 6 M HCl, 1 M HCl, and 2 M HNO_3.

Quantitative analysis is by photometry, except for uranium which is determined by fluorometry [277].

On-line procedures

Column chromatography coupled with direct detection is becoming increasingly important. The number of utilizable detectors is no longer limited to the tried UV/VIS flow-through detector. AAS and electrochemical methods have meanwhile proved their value for specific analytical problems in combination with HPLC.

Cr: Organometallic compounds can be separated on Porasil A silica gel with 0.5% of pyridine in toluene and detected, sensitively and selectively, by AAS [195].

Cu: Ionic complexes with ethylenedinitrilotetraacetic acid, nitrilotriacetic acid, ethylenebis(oxyethylenenitrilo)tetraacetic acid, and 1,2-cyclohexylenedinitrilotetraacetic acid can be separated on an anion exchanger and likewise detected by AAS [196, 197].

Sb: Apart from determinations in alloys (see Section III/1.3), this procedure is also suitable for trace analysis in hair at contents of 1 ppm [409].

Pt: The determination of the platinum compound cis-dichlorodiammineplatinum(II), used in medicinal chemistry, is performed by transformation, after chromatography, into the photochemically measurable carbamate by admixture of the reagent [41].

S: Elemental sulfur can be separated from organic compounds on Poragel 500 with various organic solvents such as chloroform or tetrahydrofuran and with mixtures of methanol and these solvents; reversed phase effects play a part in this separation [64].

Se, NO_2^-: Selenium can be separated, as in gas chromatography, as piazselenole (as 5-chloropiazselenole [399]; as naphthylpiazselenole [440]) by HPLC; nitrite can be converted by naphthyldiamine into 2,3-naphthotriazole [440].

NO_2^-, NO_3^-: After separation on a strongly basic anion exchanger these anions are determined with a Cd-coulometric flow-through detectors [81].

Zn: A reaction detector with zincon using an air-segmented flow system has been employed for detection of zinc [281]. Quantitative analyses are possible down to 2 ppb after enrichment on ion exchangers. Adsorption is from an acetic acid solution, and elution is with 0.1 M NH_4Cl solution.

FPO_4^{2-}: Combination of an ion exchange column with a Technicon Auto-Analyzer permits quantitative analysis of this special anion in one analytical step. Substances interfering with photometric determination as molybdenum blue are removed from the phosphate ions produced on hydrolysis by dialysis in the Auto-Analyzer [24]. A relative standard deviation of 3.4% is calculated for contents of 0.09%.

Cl^-, Br^-, I^-: The anions are separated on Zeokarb 225 ion exchanger (in the Cd form) and determined potentiometrically by means of a silver/silver chloride microelectrode [104].

Polythionates can be quantitatively assayed by UV photometry or after reaction with cerium(IV) salt solutions (to give fluorescing cerium(III) ions) in a reaction detector [442].

Anions and cations: Use of so-called ion chromatography (IC), which leads by combination of two ion-exchange columns to neutralization of the eluent (see Section I/1.2.4), permits separation of numerous cations and anions in a wide variety of materials and their conductometric determination in a flow system [367].

2.4 Gas Chromatography

For some elements, which can be determined only after complicated and time consuming sample preparation or with frequently insufficient sensitivity for extreme trace analysis, gas chromatography is a practicable solution, e.g. for the elements aluminum, beryllium, chromium, mercury, and selenium.

1) Aluminum [635, 565, 649]

The trifluoroacetylacetone complex can be extracted directly from water at pH 4.5 – 5.5 into organic solvents such as toluene [635], benzene [565, 649]. Gas chromatography gives interference-free analyses from water or digestion solutions of biological materials. The relative standard deviations are about 10% for concentrations of 10 ng/cm^3 water [565].

Table 59

Chromatographic analysis of organic, biological, and special environmental materials – gas chromatography.

Element	Material	Compound/ Detector	Detection limit	Ref.
	Water			
Al		$Al(TFA)_3$/ECD	1 ppm	[635]
As		As diethyldithio-carbamate/ECD	1 ppm	[532]
Cl		Phenyl-Hg chloride/FID	8 ppb	[481]
Cr		$Cr(TFA)_3$/ECD	0.1 $\mu g/dm^3$	[637]
F		$(C_2H_5)_3SiF$/FID	30 ppb	[495, 496]
Hg		Dialkyl-Hg/ECD	10 pg Hg absolute	[712]
Hg		CH_3HgX/ECD	2.5 ppb	[810]
Hg		Phenyl-Hg-X/ECD	20 ppb	[651]
P		Elemental P/FPD	ppb range, 1 pg absolute	[457]
P		Elemental P/FPD	0.2 ppb	[454]
P		Trimethylsilyl phosphate/FID, FPD	0.1 ppm	[645]
Se		Dibromopiaz-selenole/ECD	2 ng/dm^3	[728]
Se		5-Nitropiaz-selenole/ECD	20 ppb	[724]
Se		Piazselenoles/MED	0.1 ppb	[762]
As, Sb		Triphenyl cpds./ MED	50 and 30 ppb	[764]
Al, Cr		TFA chelates/ECD	4 and 0.5 ppb	[565]
Cu, Ni		Bis(acetylpivaloyl-methane)ethy-lenediimine chelates/FID		[487]
CN^-		CNBr/ECD	ppb range	[655]
CN^-, SCN^-		CNBr/ECD	10 ppb	[654, 655]
NO_3^-, NO_2^-		Nitrobenzene/ECD	0.1 ppm	[563]
	Air			
Be		$Be(TFA)_2$/ECD	0.1 ng absolute	[656]
CO		CH_4/FID	1 ppm	[774]
CO				[756]
CO		CO/WLD	1 ppm	[754]
CO		CO/Helium detector	50 ppb	[504]

Table 59 (Continued)

Element	Material	Compound/ Detector	Detection limit	Ref.
S		SO_2/FPD	5 ppb	[505]
S		SO_2/Fluorescence detector	0.2 ppm	[658]
S		H_2S, SO_2/FPD	ppb range	[666]
N		NH_3/TCD	100 ppm	[679]
	Biological material			
Al	Rat liver	$Al(TFA)_3$/ECD	ppb range	[649]
As	Urine	As diethyldithio-carbamate/ECD	1 ppm	[532]
As	Organ samples	TMS arsenate/FID	0.1 ppm	[631]
As	Liver	Triphenylarsane/ FID	0.25 ppm	[752]
Be		$Be(TFA)_2$/ECD	10 ppb	[610]
Be	Urine	$Be(TFA)_2$/ECD	1 ppb	[550]
Be		$Be(TFA)_2$/ECD	ppm range	[551]
Be	Blood, tissue	$Be(TFA)_2$/ECD	20 ppb	[772]
Be	Body fluids	$Be(TFA)_2$/ECD	2 ppb	[773]
Br	Blood	1,2-Dibromocyclo-hexane/FID	0.1 $\mu g/cm^3$	[462, 798]
Cr	Urine	$Cr(TFA)_3$/FPD	10 ppb	[687]
Cr	Urine	$Cr(TFA)_3$/ECD	3 ppb	[703]
Cr	Plasma	$Cr(TFA)_3$/MED	5 ppb	[493]
Cr	Liver samples	$Cr(TFA)_3$/ECD	20 ppb	[498]
Cr	Blood, plasma	$Cr(TFA)_3$/ECD	5 ppb	[574]
Cr		$Cr(TFA)_3$/ECD	ppb range	[709, 711]
Cr	Serum	$Cr(TFA)_3$/ECD	10 ppb	[710]
F	Urine, serum, bone	$(CH_3)_3SiF$/FID	10 ppb	[552]
F	Tissue	$(C_2H_5)_3SiF$/FID	1 ppm	[700]
Hg	Fish	CH_3HgCl/ECD	0.1 ppm	[797]
Hg	Fish	Alkyl-Hg cpds./ AAS (flame)	0.3 ppm	[516]
Hg	Fish	CH_3HgCl/ECD	2.5 ppb	[606]
Hg	Tissue	Phenyl-Hg chloride/ECD	ppb range	[638]
Hg	Urine, serum	Phenyl-Hg chloride/ECD	20 ppb	[651]
Hg		CH_3HgX/ECD	1 ppb	[518]
P	Nucleotides	Trimethylsilyl phosphate/FID	μg range	[576-578]
P		Elemental/FPD	ppb range	[457]

Table 59 (Continued)

Element	Material	Compound/ Detector	Detection limit	Ref.
Se	Liver	Piazselenoles/ ECD	1 pg absolute	[726]
Se		Piazselenoles/ ECD	10 ppb	[758]
Se		Piazselenoles/ ECD	5 ppb	[673]
Se	Urine	Piazselenoles/ ECD	7 ppb	[808]
CN^-		CNCl/ECD	25 ppb	[793]
NO_3^-, NO_2^-	Blood	Nitrobenzene/ECD	0.1 ppm	[775]
	Foodstuffs			
I	Milk	Iodoacetone/ECD	3 ppb	[569]
Ni	Tea, fats	Ni-TTFA chelate/ ECD	ppb range	[474]
P	Fruit juice, wine	Trimethylsilyl phosphate	ppm range	[702]
Se	Milk, milk products, albumin	5-Nitropiaz-selenole/ECD	5 ppb	[725]
	Miscellaneous materials			
As	Soil samples	AsH_3/MED	0.2 ppm	[705]
Be	Orchard leaves (NBS-samples)	$Be(TFA)_2ECD$	pg range abs.	[492]
Ge	Coal	$GeCl_4$/TCD	3.3 ppm	[712]
F	Toothpaste	$(CH_3)_3SiF$/FID	0.1%	[528]
Hg	Sediments	CH_3HgCl/ECD	2.5 ppb	[606]
Cu, Ni, Zn	Sediments	Diethyldithio-carbamates/FID	1 ppm	[677]
H_2O	Organic liquids	Acetylene/Rever-sion GC: TCD	0.2 ppm	[611]
HCN	Exhaust gases	HCN/FID	1 ppm	[652]
H_2O	Organic solvents	H_2O/TCD	2 ppm	[707]
H_2O	Organic solvents	H_2O/TCD	ppm range	[525]

2) Chromium [565, 637, 687, 703, 493, 498, 574, 709, 711, 710]

Chromium(III) trifluoroacetylacetonate is not formed below 55 – 60 °C; this complex can be extracted in the same way as that of Al and determined alongside the

Al chelate by gas chromatography [565]. Moreover, cis/trans isomers occur; they can also be separated [489] but this is not necessary in quantitative analysis of chromium.

This procedure is used primarily for determining physiological contents in biological materials – serum [709, 574, 711, 710, 493], urine [711, 687, 703], and tissue [498]. Apart from the ECD, the flame photometric detector [687] and the MED [493] are used.

The recoveries from urine and serum are reported, depending upon the method of sample preparation, as 75% [709], 79 – 92% [687], 86.4 – 94.6% [493], and 97.3% [637]. Sample volumes of 50 μl are sufficient for analysis of urine and plasma; standard deviations of 2% can be attained for contents of 5 ng/cm^3 [574].

At physiological levels, standard deviations of below 10% are given for all procedures: 1.4% at 0.1 μg/dm^3 in water [637]; 8% at 60 ppb in urine [687].

Comparison with AAS demonstrates the higher sensitivity of the GC-ECD method [711].

3. Beryllium [550, 551, 772, 773, 656]

This element can also be extracted as the trifluoroacetylacetonate with benzene and thus enriched for GC analysis.

Sample preparation with excited oxygen (microwave field) and with nitric acid/hydrofluoric acid in a Teflon vessel under pressure have been described [610].

Interfering aluminum can be precipated as the oxinate [656]. Very low volumes of 50 μl can again be used for this determination [773]. Recovery is reported as 92% for blood or 69% for liver tissue [772], and 70 – 90% for various other materials [656]. The standard deviations in the ppm to ppb range are comparable with those of chromium determination (vide supra).

4. Mercury [721, 810, 651, 797, 516, 606, 638, 518]

Gas chromatography represents the most important method available for analysis of mercury in biological and environmental samples. It is able to separate organomercury compounds and provides for sensitive detection by an ECD or element-selective MED and

AAS detectors. Examples of analysis of **organomercurials** in various materials are to be found in the publications [721, 810, 797, 516, and 518]. The effort invested in sample preparation can be greatly reduced by use of AAS or MED for detection and quantitation.

Mercury ions can also be converted in aqueous solution by various reagents (see Section I/1.2.2 – Table 19) into methyl and phenyl-mercury compounds which, after extraction and enrichment, can be determined by gas chromatography [810, 651, 606, 638, 518]. The reagents used were tetraphenylboron [810, 638], lithium pentafluorobenzenesulfinate [651], sodium 4,4-dimethyl-4-silapentane-1-sulfonate [606], and tetramethyltin [518]. Adsorption effects in packed columns necessitate particularly careful conditioning of the columns for quantitative analysis in the extreme trace region (see e.g. [638]). Alkyl- and aryl-mercury halides exhibit more favorable gas-chromatographic properties than diorganyl compounds. For sample, diphenylmercury is transformed into PhMgBr by treatment with $HgBr_2$ before gas chromatography [721]. Moreover, chloride contents can be determined as PhHgCl after reaction with diphenylmercury [481]. The standard deviations for 8, 20, and 60 ng/cm^3 of the element in water are 29, 7.2, and 4.5%, respectively [481].

At 1.6 ppm levels of mercury in water, urine, and serum, recoveries of 70.5, 81.4, and 51.0% were obtained with relative standard deviations of 6.8, 10.3, and 9.4% after an analysis time of eight minutes for chromatography [651].

Reaction with tetraphenylboron in the case of serum and liver samples gave recoveries of 80.4 ± 8.7% and 93.4 ± 8.3% at contents of 3.6 ppm [810].

5. Selenium [728, 724, 762, 726, 758, 673, 808, 725]

After reaction with aromatic diamines to give **piazselenoles**, selenium(IV) ions in aqueous solutions can be determined by photometry or fluorometry. Determination of these compounds by gas chromatography (i) largely avoids the extensive interference attending the spectroscopic methods and (ii) permits a considerable increase in sensitivity by use of an ECD with halo- and nitrosubstituted diamines.

The following derivatives have been used for determinations in biological and environmental materials: 5-nitropiazselenole [725, 673, 762, 724], 4,6-dibromopiazselenole [728], 4,5-dichloropiazselenole [758], naphthylpiazselenole [808].

The materials are wet ashed, and the selenium is reduced to Se(IV) and treated with the corresponding diamines. Extraction of the resulting piazselenole into benzene or toluene leads to enrichment prior to gas chromatographic analysis. An ECD is generally employed; high sensitivity is also observed with an MED [762].

The standard deviations are 3% for 0.98 ppm, and about 5% for 83 ppb [762]. The same order of magnitude of standard deviation is observed for GC analyses with an ECD. Various sample materials require preparation procedures of varying complexity; direct reaction and extraction of the resulting piazselenole is possible with aqueous samples.

As: Inorganic and methylated arsenic compounds can be separated by gas chromatography after reaction with sodium diethyldithiocarbamate; careful column conditioning is necessary. A recovery of 82.6% with a standard deviation of 9.5% was found for 1 ppm of inorganic arsenic [532].

Arsenic has also been determined as the **trimethylsilyl ester** of **arsenic acid** [631]. Acid treatment in a Teflon vessel is followed by extraction as diethylthiocarbamate, renewed extraction as arsenic acid, and subsequent silylation. A standard deviation of 15.7% is reported for 1.4 ppm in liver [631]. Dried samples can be incinerated in a Schöniger flask for GC determination as **triphenylarsane** [752]. The arsenic is again extracted as carbamate before reaction with diphenylmagnesium.

As, Sb: The two elements are preconcentrated by precipitation with thionalide and treated with phenylmagnesium bromide. Analyses of aqueous samples having contents of 0.5 ppb are possible with a standard deviation of 6% [764].

F: On GC determination of this anion as triethylfluorosilane with a FID a standard deviation of 6% is reported for concentrations of 0.95 $\mu g/cm^3$ [495, 496].

Recovery from samples of liver after acid extraction in a distillation apparatus is 82 – 93%; a standard deviation of 35% was found for contents of 90 ppb [700].

In two other procedures, the fluoride is extracted as trimethylfluorosilane with benzene [552, 528].

P: Elemental phosphorus can conveniently be separated by gas chromatography as phosphorus hydride and determined without interference with a flame photometer at 526 nm [457, 454].

Phosphate ions are first transformed into tristrimethylsilyl phosphate, which is also suitable for detection with a FID [645]. Analysis by gas chromatography requires 12 min.

Cu, Ni: A suitable chelating agent is bis(acetylpivaloylmethane) ethylenediimine; the complexes can also be determined by photometry. A standard deviation of 6% is found for 0.25 $\mu g/cm^3$ of copper in water [487].

CN^-, SCN^-: Reaction with bromine affords derivatives which can be separated by gas chromatography on Porapak and sensitively determined with an ECD [654, 655]. Cyanide can be determined in various biological materials as cyanogen chloride by reaction with chloramine-T. Comparison with a photometric method gave results is good agreement [793].

NO_3^-, NO_2^-: Recovery as nitrobenzene was 90 ± 8% for 0.12 – 62 ppm of nitrate, and 92 ± 12% for 0.09 – 37 ppm of nitrite (after oxidation with potassium permanganate) [563].

Recovery of nitrate and nitrite traces after reaction with trimethoxybenzene is 81 ± 3% [775].

Inorganic gases are determined in an extremely wide range of materials.

In a gaseous matrix: Air, respiratory air, exhaust gases, combustion gases, pyrolysis gases, hydrocarbons.

In a liquid matrix: Water, blood, organic solvents.

In a solid matrix: Coal (see [701]).

Work prior to 1970 is cited in the publication by *Rüssel* and *Tölg* [701].

Recent work with TCD and FID's, as well as a helium detector, a flame photometric, and a special fluorometric detector, is listed under the heading "Air" in Table 59 (for carbon monoxide and sulfur-containing gases).

Br^-: Low bromide contents in blood can be extracted completely as 1,2-dibromocyclohexane into cyclohexane [462]. Another procedure utilizes reaction with 2,3-dimethylphenol to give brominated derivatives [798].

I^-: The standard deviation for GC analysis of 1 ppb of inorganic iodine as iodoacetone is 16 – 20%, and less than 10% for 36 ppb of iodide [569].

Ni: Nickel can be extracted as monothiotrifluoroacetylacetonate at pH 4.5 – 5.0; the recovery for 0.1 $\mu g/cm^3$ is 94 ± 7%. Analysis at levels of 13 ppb in tea give a relative standard deviation of 10.8% [474].

Inorganic compounds of, e.g., arsenic and germanium can be determined by gas chromatography:

As is first extracted from the sample by treatment with nitric acid and then converted by reduction with KI, $SnCl_2$, and zinc into arsenic hydride [705].

Ge is isolated as tetrachloride by chlorination from coal [712]. The standard deviation is 6% for 3.3 ppm Ge.

Cu, Ni, Zn: After acid treatment, and separation from iron and manganese by precipitation with $La(OH)_3$, these elements can be extracted as diethylthiocarbamates and determined in this form by gas chromatography [677]. Analytical results of 1.6, 4.9, and 17.1 ppm have been obtained for Cu, Ni, and Zn in marine sediments.

H_2O: Water can be determined directly with a TCD [707] or after reaction to form acetylene (ethyne) [611] with a FID or a gas density balance [525].

Sephadex [525], Porapak Q [707], or a molecular sieve [611] serves as separation material.

Enrichment by reversion gas chromatography leads to detection limits in the ppb range [611].

3 Chromatographic Enrichment Methods

In **liquid chromatography**, a higher concentration than that in the analytical sample can be attained in the eluate whenever the mobile phase of the adsorption step differs from that of the elution process. Enrichment is possible in **gas chromatography** if adsorption is performed at low column temperatures and separation and elution are conducted with the aid of an increasing temperature gradient.

The modest number of examples listed in Table 60 provide a survey of the various chromatographic (broadly speaking) methods available for enrichment in inorganic analysis. General **adsorbents** such as active charcoal are suitable for liquids [427] and molecular sieves for gases [131].

Ion exchange chromatography is of widespread use as a method for enriching cations and anions: in addition to cation and anion exchange resins, increasingly frequent use has recently been made of chelate exchangers as a selective and powerful means of enrichment of traces of elements. The chelating groups are chemically bonded to the matrix e. g. by azo coupling, and generally consist of reagents which have already proved their worth in the photometric determination or chromatography of metal chelates. Suitable matrices are polymers, cellulose, and glass. Stable materials may also be formed in the absence of a strong chemical bond, as with polyethylene imine-cellulose [453] or polymeric (polyurethane) foam with chelating agents [42 – 44], which are particularly suitable for trace enrichment of individual elements.

Other special methods from the field of ion-exchange chromatography are **reactive adsorption** of oxygen-containing anions [260] and modern ion chromatography (IC) with special concentration columns [438].

Reversed phase methods are also widely employed for separation and enrichment of elements. Three

Table 60
Trace enrichment by chromatographic methods (selection).

Elements	Method: Chromatographic system	Material	Enrichment, detection limit	Ref.
	1) *Adsorption*			
Cations	Active charcoal LC	Water	10^4	[427]
CO, permanent gases	Molecular sieve GC	Air	0.01 ppm – TCD	[131]
	2) *Ion exchange – LC*			
	a) *Anion exchangers*			
U	Uranyl thiocyanate complex	Water	0.04 µg/dm³ – NAA	[55]
Co, Cd, Zn	Complexes with 2-(3-sulfobenzoyl)pyridine 2-pyridylhydrazone	–	300	[135]
Cr, Pb	Aqueous solutions	Mixtures	10^{-6} mol/dm³ – Volum.	[312]
	b) *Cation exchangers*			
Au	Dowex 1 × 8	–	–	[388]
24 Cations	Carboxycellulose	–	1 – 100 µg/dm³	[389]
Tl	CE	Seawater	0.1 ppm – F, NAA	[282]
15 Elements	Dowex	Silicates	ppm range – AAS	[284]
Pt Elements	CE	Metals	ppm range	[332]
	c) *Chelate exchangers (ChA) (with chemically bonded ligands)*			
Cd, Co, Cu, Ni, Zn	Chelex 100	Natural waters	µg range/dm³	[98]
Heavy metals	Chelex 100	Geological samples	ppm range – AAS photometry	[209]
Heavy metals	Chelex 100	Seawater	ppm range	[99b, 245]
Numerous cations	Ion exchange membrane	Water	1250	[428]
Cr, V	As peroxy cpds; adsorption on ChA	–	1 µg – Photometry	[283]
Noble metals	Various chelating groups on styrenedivinylbenzene polymers	–	–	[341]
Noble metals	Copolymer of styrene and 3(5)-methylpyrazole – TLC	Solution	–	[9]

Table 60 (Continued)

Elements	Method: Chromatographic system	Material	Enrichment, detection limit	Ref.
Pb, Zn, Cd, Mn	Dithizone or oxine diazo-coupled to modified carboxymethyl cellulose	Seawater	$10^5 - 10^8$	[21]
Cu	8-Hydroxyquinoline (oxine), immobilized on glass spheres	Water	$\mu g/dm^3$	[144]
Fe, Cu, Zn Pb, U	Cellulose chelate exchanger (Hyphan)	Water	1 ppb – XFA	[58]
U	Chelating cellulose	Water	0.3 ppb – XFA	[59]
Heavy metals	Dithiocarbamate exchanger	–	ppm-ppb range	[87]
	d) *Special exchangers*			
MnO_4^-, CrO_4^{2-}, VO_3^-	Reactive adsorption on Fe(III) exchanger	–	1:40, ppb range – AAS, photometry	[260]
Cu	Polyethylene imine cellulose	Water	$1:10^5$ – photometry	[453]
Ions in general	Ion chromatography (IC)	–	–	[438]
Ag	Polyurethane foam + TBP	–	–	[42]
Ag	Polyurethane + dithizone	Water	0.01 $\mu g/dm^3$	[43]
Co	Polyurethane + 1-nitroso-2-naphthol, diethyldithiocarbamate	Water	–	[44]
	3) *Reversed phase systems*			
Ga e.g.	Teflon + TBP	–	–	[7]
			see also	[42]
Au	Fluoroplast + TBP	Alloys	–	[305]
	4) *Paper chromatography*			
Trace elements	Survey	Geochemical materials	10^4	[286]
	5) *Reversion gas chromatography*			
Inorganic gases	Survey	Air etc.	ppb range	[611a, 611b]

Abbreviations:

LC, liquid chromatography; CE, cation exchange; NAA, neutron activation analysis; F, fluorometry; XFA, X-ray fluoroscence analysis; TBP, tributyl phosphate.

Table 61
Chromatographic methods for analysis of elemental species.

Element: Species	Methods	Material	Ref.
As: Alkylarsonic acid	GC/MED	Environmental matrices	[763]
As: Methyl-As cpds./inorganic As	GC/ECD	Water, urine	[532]
As, S: Inorganic ions	IC/Conductometry	Environmental matrices	[146]
Cu: Amino acid complexes	HPLC-IE/AAS	Water	[196, 197]
Cu: Amino acid complexes	HPLC-IE/AAS	Serum	[202]
Hg: Organometallic Hg cpds.	GC, TLC (survey)	Water, biol. mat.	[99a]
Hg: Organometallic Hg cpds./inorganic Hg	GC methods (survey)	Miscellaneous	[685]
Hg: Methyl-Hg cpds.	GC/ECD	Fish	[745]
Hg: Volatile Hg-organic cpds.	GC/MED	Tissue, water	[761]
Hg: Methyl-Hg cpds.	GC/ECD	Fish	[797]
Hg: Organometallic cpds.	HPLC/Voltammetry	Biolog. mat.	[265]
Pb: Tetraalkyl-Pb cpds.	GC/AAS	Air	[522]
Pb: Tetraalkyl-Pb cpds.	GC/AAS	Water, sediments, fish	[523]
Pb: Alkyl-Pb cpds.	GC/AAS	Air	[678]
Pb: Tetraalkyl cpds.	GC/MED	Air	[680]
Pb: Organic Pb cpds.	GC/AAS (flameless)	Air	[683]
Pb: Organometallic cpds.	HPLC-Gel/AAS	–	[224]
S: SO_2, H_2S	GC/FPD	Air	[757]
Se: $(CH_3)_2Se$, $(CH_3)_2Se_2$	GC/AAS	Air	[521]
Sn: Methyl-Sn cpds.	TLC-Ph, GC	–	[625]
Sn: Phenyl-Sn cpds.	GC/FID, MS	Water	[735]
Pt: cis-Dichlorodiammine-Pt(II)	HPLC/RD-UV	Urine	[14]
Hg, Cu: Phosphate complexes	HPLC-Gel/AAS	–	[450, 451]
Hg, As, Sn: Phenyl cpds.	HPLC/AAS, flameless	–	[54]
Cu, Ni, Zn: Miscellaneous	Gel filtration	Water	[269a, 284a]
Cu, Pb, Cd, Zn: Miscellaneous	Chelate ion exchange	Water	[99b, c]
Fe, Mn, Ni, Cu, Cr, Pb, Cd: Adsorbates	Gel chromograph	Water	[399a]
Cr: Chromium(III), Chromate	HPLC-carbamates, RP/UV	Water	[394]

IC, ion chromatography; IE, ion exchange; RD, reaction detector; Ph, photometry.

examples with tributyl phosphate as stationary phase are included in Table 60.

Not only column chromatography techniques, often in batch operation, but also **layer chromatography**, e. g. in the form of paper chromatography [286] or as special thin-layer chromatography [9], have been used as enrichment methods in inorganic trace analysis.

In **gas chromatography** the **reversion technique** has secured an firm place for temperature-dependent enrichment and subsequent separation of numerous inorganic gases in the ppb to ppt range (see, e. g. [611 a, b]).

Whenever it could be seen from the reference, the enrichment factor (e. g. 10^4, 300, or $1:10^5$) is given under the heading "Enrichment, detection limit"; in other cases the detection limit is listed, which will obviously depend upon the method of determination employed.

4 Chromatographic Analysis of Element Species

The toxicity of heavy metals in biological systems is crucially dependent upon the form in which they are chemically bonded, designated as "species".

For example, in natural samples of water chemical elements may be present in solution in ionic form, as unionized chelates, or as organometallic compounds at trace levels.

Chromatographic methods are very important for separation and determination of such element species owing to their great variety. Differential analysis of **organomercurials** would be unthinkable without gas chromatography and numerous publications applying to a wide range of materials demonstrate the performance of this method (see Table 61).

The analysis of organometallic compounds is also of interest for **arsenic, lead, selenium**, and **tin**.

Methods of liquid chromatography are employed for separation of ionic species, with possible differentiation between simple and complex ions.

Methods of gel filtration and gel chromatography have been used for studies on the distribution of elements on suspended or colloidal inorganic or organic materials (see [269a, 284a, 399a]).

A survey of chromatographic methods for this increasingly important area of inorganic trace analysis is given by Table 61.

5 Comparison of Chromatographic Analytical Methods with Those of Spectroscopy and Electrochemistry

Extreme trace analysis of elements ventures into regions of concentrations and quantities in which numerous sources of interference are to be expected which greatly falsify the results. In order to answer the resulting question of **accuracy**, analyses are performed by various methods and results showing good agreement are generally to be regarded as realiable or correct.

Chromatography in particular can make an important contribution in this context because it often represents a method which is completely independent of the usual spectroscopic and electrochemical methods of element analysis.

Comparative analysis with methods of photometry, fluorometry, emission spectrometry, atomic adsorption spectrometry, X-ray fluorescence analysis, neutron activation analysis, and voltammetry/polarography have been published for the elements aluminum, beryllium, bromine, chromium, germanium,

Table 62
Comparison of methods for trace analysis of elements.

Element (Material)	Methods	Results	Ref.
Al (Uranium)	GC: TFA chelate/ECD	11.6 ppm	
	Photometry (extractive + 8-hydroxyquinoline)	14.5 ppm	[561]
Be (Granite)	GC: TFA chelate/ECD	2.07 ± 0.13 ppm	
	Spectrochemical	2.3	
	Fluorometric	3.20 ± 0.03	
(Basalt)	GC (as above)	1.53 ± 0.18	
	Spectrochemical	2	
	Fluorometric	1.59 ± 0.11	
GC and fluorometry show about the same sensitivity			[540]

Table 62 (Continued)

Element (Material)	Methods	Results	Ref.
Be (Orchard leaves – Standard from NBS, USA)	GC (as above)	Not determinable (interference in chromatogram)	
	Emission spectrometry	Detection limit 0.02 ppm	
	Fluorometry (+ morin)	0.0213 ± 0.0035 ppm	[546]
Be (various environmental materials	GC-ECD	1) 5.2×10^{-12} g/cm^3 2) 7.2×10^{-12} 3) 8.5×10^{-12}	
	GC-MS	1) 5.6×10^{-12} 2) 7.7×10^{-12} 3) 8.9×10^{-12}	
(Also compares various wet and dry methods of sample preparation.)			[492]
Br (Bromide) (Blood)	GC-FID (as 1,2-dibromocyclohexane)	1) 86.2 mg/100 cm^3 2) 58.8 3) 34.3 4) 12.5 5) 7.8	
	X-ray fluorescence (XFA)	1) 84 2) 61 3) 34 4) 16* 5) 10**	
* Sample contains carbromal, ** Sample contains bromvaletone (established by thin-layer chromatography)			[462]
Cr (Urine, serum)	GC-ECD (TFA chelate)	1 ng/100 ml benzene extract	
	AAS (flame)	0.6 µg/100 ml benzene extract	
Correlation coefficient (r):		Urine: r = 0.963 Serum: r = 0.923	[711]
Cr (Urine)	GC-ECD (as above)	2.7 ± 0.36 ppb (N = 3)	
	NAA (with separation)	2.9 ± 0.44 (N = 10)	
	AAS (furnace)	2.66 ± 0.12 (N = 3)	
	GC-ECD	41.4 ± 2.20 (N = 4)	
	NAA	49.5 ± 3.9 (N = 8)	
	Flame emission	47.0 ± 5.4 (N = 3)	
	AAS (furnace) 1000 °C	51.7 ± 2.08 (N = 3)	
	700 °C	60.5 ± 0.58 (N = 3)	
(Results from various laboratories are compared.)			[703]

Table 62 (Continued)

Element (Material)	Methods	Results	Ref.
CN^- (Cyanide) (biological material) – after poisoning (from corpses)	GC-ECD (as CNCl)	1) Blood 4.3 $\mu g/cm^3$ 2) Blood 60 3) Gastric fluid 4100	
	Photometry	1) 4.0 2) 53 3) 3900	[793]
Ge (Coal)	GC-TCD (as $GeCl_4$) Photometry (+ phenylfluorone)	Mean values in ppm range differ by about 7%	[712]
Hg (Sediments, fish)	GC-ECD (as CH_3HgX)	Sediment: 44.0 – 45.5 – 47; 136 – 132 – 129 ppb Fish: 2.79, 2.89, 2.84; 8.58, 8.56, 8.53 ppb	
	AAS (cold vapor method)	Sediment: 44.1 ± 9.13 ppb; 109.6 ± 21.5 ppb Fish: 2.06 ± 0.83; 7.23 ± 2.02 ppb	[606]
Ni (Tea)	GC-ECD (thiotrifluoroacetylacetonate)	13.0 ppb (s_r: 10.8%)	
	UV photometric (derivative as above)	14.0 ppb (20.0%)	
	AAS (flame)	12.7 ppb (19.7%)	[474]
N (Nitrate) (Cereal stalks)	LC/coulometry (ion exchange)	1) 130 ppm 2) 770 3) 860	
	Photometric	1) 160 2) 700 3) 840	[81]
Se (Water)	GC-MED (piazselenole)	Detection limits: 0.1 ppb	[762]
	HPLC-RP/UV-320 nm (5-chloropiazselenole)	0.32 ppb	[399]
	Photometry	5.3 ppb	[399]
Se (Aqueous solutions)	GC-ECD (5-nitropiazselenole)	Detection limits in g: 10^{-11}	
	NAA	10^{-9}	
	AAS (flameless)	7×10^{-11}	
	Voltammetry	4×10^{-9}	
	Fluorometry	2×10^{-9}	
	Photometry	2×10^{-8} to 2×10^{-9}	[392]

Table 62 (Continued)

Element (Material)	Methods	Results	Ref.
Se (Serum, plasma)	GC-ECD (naphthylpiaz-selenole)	1) 0.52 μg/cm^3 2) 0.40 3) 0.36	
	NAA	1) 0.44 2) 0.46	
(various samples)		3) 0.39	[808]
Sn (Alloy: zircaloy)	GC-TCD ($SnCl_4$)	1.30 ± 0.03%	
	XFA	1.40 ± 0.04	
	Polarography	1.37 – 1.38	[476]
Cations and anions (Water) **Na**	Ion chromatography (IC)-Conductometry	22; 49 ppm	
	AAS	23; 48	
K	IC	1.4; 4.0 ppm	
	AAS	2.2; 3.7	
Ca	IC	43; 74 ppm	
	AAS	47; 74	
	Complexom.	46; 74	
Mg	IC	13; 21 ppm	
	AAS	12; 19	
	Complexom.	14; 20	
Cl^-	IC	46; 122 ppm	
	Argentometry	54; 117	
	NAA	50; 118	
SO_4^{2-}	IC	28; 44 ppm	
	Turbidimetry	30; 49	[367]

TFA, trifluoroacetylacetone; MS, mass spectrometer; NAA, neutron activation analysis; LC, liquid chromatography; RP, reversed phase chromatography; IC, ion chromatography.

mercury, nickel, selenium, tin, etc. in a number of important papers. The results obtained with regard to mean values of analyses in various materials, the relative standard deviations and reproducibility criterion, and a comparison of performance on the basis of detection limits are to be found in Table 62.

IV References

The choice of original papers evaluated from the thousands of publications concerned with the chromatography of inorganic substances was dictated primarily by the two principal objectives of this book:

- to survey the possibilities inherent in chromatography for the separation and determination of inorganic materials, and
- to list the applications of chromatographic analysis to inorganic trace analysis.

Older work (before 1970) not dealing specifically with problem-oriented applications has therefore been included in part only in order to present an overview. Work published since 1970 has been covered almost exhaustively; particularly that concerned with quantitative trace analysis.

In the case of publications accessible only with difficulty, such as those appearing in Russian, Asian, and particularly eastern European journals that are not readily available in the western hemisphere, and are not published as English translations, the corresponding abstracts in *Chemical Abstracts, Analytical Abstracts*, and the Abstracts Section in *Fresenius' Zeitschrift für Analytische Chemie* have been evaluated and cited as the only source (marked with an asterisk).

The titles of the publications cited are given in full, often providing the reader with valuable additional information going beyond the immediate context of the citation. The original titles are given for English, French, and German publications, and English translations are given for papers in other languages.

Further sources of information about publications on the chromatography of Inorganic Substances are the periodical literature surveys in *Journal of Chromatography* and the methodological reviews in *Analytical Chemistry*.

The abstracts journals are abbreviated as C.A. for *Chemical Abstracts* and Anal. Abstr. for *Analytical Abstracts.*

English translations of journal titles are *Japan. Analyst* for Bunseki Kagaku and *Russian Journal of Analytical Chemistry* for Zh. Anal. Khim.

Liquid Chromatography

[1] *S. Abe* and *H. Weisz,* "The rapid separation and determination of uranium(VI) by use of DEAE anion-exchange paper and the ring-oven technique", Mikrochim. Acta (1970) 550.

[2] *L. H. Adcock* and *W. G. Hope,* "A method for the determination of tin in the range 0.2 – 1.6 μg and its application to the determination of organotin stabiliser in certain foodstuffs", Analyst **95** (1970) 868.

[3] *H. Agrinier,* "Application of paper chromatography for the determination of trace elements in geological materials", C. A. **75** (1971) 104 750 j.

[4] *G. Alberti, G. Giammari,* and *G. Grassini-Strazza,* "Chromatographic behaviour of inorganic ions on crystalline titanium phosphate or zirconium phosphate thin layers", J. Chromatogr. **28** (1967) 118.

[5] *G. Alberti, M. A. Massucci,* and *E. Torracca,* "Crystalline insoluble salts of polybasic metals IV. Chromatography of inorganic ions on cerium (IV) phosphate sheets", J. Chromatogr. **30** (1967) 579.

[6] *V. N. Aleskovskaya* and *V. B. Aleskovskii,* "Determination of micro-amounts of certain cations by peak precipitation chromatography on paper impregnated with slightly soluble diethyldithiocarbamates", Zh. Anal. Khim. **25** (1970) 243.

[7] *I. P.Alimarin* and *T. A. Bolshova,* "Separation and concentration of elements by reversed-phase partition chromatography", Pure Appl. Chem. **31** (1972) 493.

[8] *B. Andreasen et al.,* "Slip chromatography: a semi-quantitative TLC technique", J. Chem. Educ. **53**, (1976) 772.

[9] *I. I. Autokol'Skaya et al.,* "Concentration and separation of elements on chelate sorbents. A sorbent for the noble metals based on the copolymer of styrene and 3(5)-methylpyrazole", Zh. Anal. Khim. **31** (1976) 742.

[10] *S. Araki et al.,* "Automated ion-exchange chromatography for the ultra-micro determination of alkali and alkaline-earth metals. Use of a double-jet hydrogen flame ionisation detector", Bunseki Kagaku **19** (1970) 493.

[11] *S. Araki et al.*, "Automatic ion-exchange chromatography for analysis of alkali-metal and alkaline-earth metal mixtures", Talanta **19** (1972) 577.

[12] *J. Bäumler* and *S. Rippstein*, "Zur Mikroquecksilberbestimmung", Mitt. Gebiete Lebensm.-Hyg. **54** (1963) 472.

[13] *K. Ballschmiter*, "Chromatographie von Metallchelaten I. Dünnschicht-chromatographische Trennung substituierter Dithiocarbamidate: Chelate des Diäthyldithiocarbamidats", Z. Anal. Chem. **254** (1971) 348.

[14] *S. J. Bannister, L. A. Sternsol*, and *A. J. Repta*, "Urine analysis of platinum species derived from cis-dichlorodiammoniumplatinum(II) by high-performance liquid chromatography following derivatization with sodium diethyldithiocarbamate" J. Chromatogr. **173** (1979) 333.

[15] *R. Baranowski et al.*, "2,2-Diquinoxalyl as a new reagent for determination of bivalent tin in thin-layer chromatography", Microchem. J. **20** (1975) 1.

[16] *L. S. Bark* and *G. Duncan*, "The chromatographic process in a reversed-phase system for the separation of some metal ions, by ion-association distribution", J. Chromatogr. **49** (1970) 278.

[17] *L. S. Bark, G. Duncan*, and *R. J. T. Graham*, "Inorganic thin layer chromatography Part II. Chromatography of some first row transistion metals on thin layers of substrates impregnated with tributyl phosphate", Analyst **92** (1967) 31.

[18] *L. S. Bark et al.*, "The chromatographic process in some reversed-phase systems used for the separation of metal ions", Symp. Chromatogr. Electrophorese VI, Bruxelles 1970, (1971) 382.

[19] *R. K. Barna* and *N. K. Baishya*, "Spectrophotometric determination of lead, copper, and bismuth in a mixture after separation by paper chromatography", Current Sci. (India) **37** (1968) 434.

[20] *T. P. Baudot et al.*, "Identification of toxic metals after extraction and thin-layer chromatography of their dithizonates. Toxicological applications", J. Chromatogr. **128** (1976) 141.

[21] *A. J. Bauman, H. H. Weetall*, and *N. Weliky*, "Coupled liquid chromatography, applications to trace element collection and characterization", Anal. Chem. **39** (1967) 932.

[22] *K. Becherer*, "Quantitative paper chromatographic determination of alkali metals", Oesterr. Akad. Wiss. Math. Naturw. Kl., Sitzber. Act. I, **175**, No. 4 – 6, (1966) 107; C. A. **69** (1968) 15 684 z.

[23] *L. Ben-Dor* and *E. Jungreis*, "Ultramicro spectrofluorimetric determination of aluminium", Isr. J. Chem. **8** (1970) 951.

[24] *C. Benz* and *R. M. Kelley*, "An automated ion-exchange method for the determination of sodium monofluorophosphate in dentifrices", Anal. Chim. Acta **46** (1969) 83.

[25] *A. Beszterda*, "Paper-chromatographic determination of microgram amounts of sulphates", Chem. Anal. (Warsaw) **14** (1969) 341; Anal. Abstr. **19** (1970) 195.

[26] *R. P. Bhatnagar* and *K. N. Bhattacharya*, "Quantitation of uranyl ion by paper chromatographic technique", J. Indian Chem. Soc. **53** (1976) 931.

*[27] *D. Biala*, "Determination of lead in the blood by thin-layer chromatography without mineralization", C. A. **81** (1974) 58 838 t.

*[28] *S. Billinski, J. Klimek*, and *D. Misiuna*, "Methodical investigations on the usefulness of 2-hydrazinothiazole derivatives for detecting microgram amounts of metal ions on paper chromatograms II. Determination of Mn(II), Co(II), Ni(II) ions by direct planimetry of spots", C. A. **76** (1972) 148 452 p.

[29] *B. Björkqvist* and *H. Toivonen*, "Sensitive high-performance liquid chromatographic method for the determination of water in various samples", J. Chromatogr. **178** (1979) 271.

[30] *E. Blasius* and *P. G. Maurer*, "Austauscher mit Monobenzokryptand K(2_B.2.2.) als Ankergruppe", J. Chromatogr. **125** (1976) 511.

[31] *E. Blasius, K.-P. Janzen*, and *W. Neumann*, "Anwendung von Austauschern mit cyklischen Polyäthern als Ankergruppe in der Mikro- und Spurenanalyse", Mikrochim. Acta II (1977) 279.

[32] *E. Blasius et al.*, "Herstellung, Charakterisierung und Anwendung komplexbildender Austauscher mit Kronenverbindungen oder Kryptanden als Ankergruppen", Z. Anal. Chem. **284** (1977) 337.

[33] *E. Blasius et al.*, "Application of exchangers with crown ethers as anchor groups in analytical and preparative chemistry", J. Chromatogr. **167** (1978) 307.

*[34] *G. Bönig*, "Determination of total strontium in vegetable material and seawater by circular paper chromatography", C. A. **68** (1968) 18 410 x.

[35] *G. Bönig* and *H. Heigener*, "Eine papier-chromatographische Abtrennung von Blei in pflanzlichem Material, seine Identifizierung und Bestimmung", Landwirtsch. Forsch. **19** (1966) 117.

[36] *G. Bönig* and *H. Heigener*, "Microdetermination of vanadium in biological substances by selective paper chromatography", Landwirtsch. Forsch. **25** (1972) 139.

[37] *G. Bönig* and *H. Heigener*, "Paper chromatographic analysis of thallium in organic material", Landwirtsch. Forsch. **26** (1973) 81.

[38] *E. G. Bogdanova* and *V. N. Aleskovskaja,* "Peak chromatographic determination of trace amounts of platinum group metals", Zh. Anal. Khim. **29** (1974) 1857.

*[39] *E. S. Boichinova* and *V. B. Aleskovskii,* "Determination of trace elements by the peak heights on paper chromatograms I. Determination of copper and nickel ions", C. A. **59** (1963) 9 308 e.

[40] *J. Borak,* "Chromatographic behaviour of inorganic salts on hydroxyethyl methacrylate gels", J. Chromatogr. **155** (1958) 69.

[41] *R. F. Borch, J. H. Markovitz,* and *M. E. Pleasants,* "A new method for the hplc analysis of Pt (II) in urine", Anal. Lett. **12** (1979) 917

[42] *T. Braun* and *A. B. Farag*, "Reversed-phase foam chromatography. Chemical enrichment and separation of gold in the tributylphosphate-thiourea-perchloric acid system", Anal. Chim. Acta **65** (1973) 115.

[43] *T. Braun* and *A. B. Farag,* "Plasticized open-cell polyurethane foam as a universal matrix for organic reagents in trace element preconcentration I. Collection of silver traces on dithizone foam", Anal. Chim. Acta **69** (1974) 85.

[44] *T. Braun* and *A. B. Farag,* "Plasticized open-cell polyurethane foam as a universal matrix for organic reagents in trace element preconcentration III. Collections of cobalt traces on 1-nitroso-2-naphthol and diethyldithiocarbamate foams", Anal. Chim. Acta **76** (1975) 107.

[45] *T. Braun, L. Bakos* and *Z. Szabo,* "Reversed-phase foam chromatography. Separation of iron from copper, cobalt and nickel in the tri-n-butyl-phosphate-hydrochloric acid system", Anal. Chim. Acta **66** (1973) 57.

[46] *U. A. Th. Brinkman* and *G. deVries,* "Liquid anion-exchangers in reversed-phase extraction chromatography", J. Chem. Educ. **49** (1972) 244.

[47] *U. A. Th. Brinkman, G. deVries,* and *E. vanDalen,* "Chromatographic technique using liquid anion-exchangers I. HCl-systems", J. Chromatogr. **22** (1966) 407.

[48] *U. A. Th. Brinkman, G. deVries,* and *E. vanDalen,* "Chromatographic techniques using liquid anion-exchangers II. Strong monobasic acid systems", J. Chromatogr. **23** (1966) 287.

[49] *U. A. Th. Brinkman, G. deVries,* and *E. vanDalen,* "The nature of some metal-chloro anions present in the organic phase in reversed-phase chromatography involving liquid anion-exchangers", J. Chromatogr. **31** (1967) 182.

[50] *U. A. Th. Brinkman, G. deVries,* and *R. Kuroda,* "Thin-layer chromatographic data for inorganic substances", J. Chromatogr. **85** (1973) 187.

[51] *U. A. Th. Brinkman, G. deVries,* and *H. R. Leene,* "Reversed-phase extraction chromatography using Br-containing eluents", J. Chromatogr. **69** (1972) 181.

[52] *U. A. Th. Brinkman, D. J. J. Sterrenburg,* and *G. deVries,* "Chromatographic techniques using liquid anion exchangers IV. HF-systems", J. Chromatogr. **54** (1971) 449.

[53] *U. A. Th. Brinkman, J. U. Taphoorn,* and *G. deVries,* "Reversed-phase extraction chromatography using a tetra-substituted pyrazole as stationary phase", J. Chromatogr. **84** (1973) 407.

[54] *F. E. Brinkman et al,* "Application of a liquid chromatograph coupled with a flameless atomic absorption detector for speciation of trace organometallic compounds", J. Chromatogr. Sci. **15** (1977) 493.

[55] *R. J. N. Brits* and *M. C. B. Smit,* "Determination of uranium in natural water by preconcentration on anion-exchange resin and delayed-neutron counting", Anal. Chem. **49** (1977) 67.

[56] *A. O. Brunfelt* and *E. Steinnes,* "Determination of lutetium, ytterbium and terbium in rocks by neutron activation and mixed solvent anion-exchange chromatography", Analyst **94** (1969) 979.

[57] *J. W. Brunstad,* "Colorimetric determination and thin-layer identification of boron as acid in caviar", J. Assoc. Off. Anal. Chem. **51** (1968) 987.

[58] *P. Burba* and *K. H. Lieser,* "Abtrennung von Spurenelementen an einem Celluloseaustauscher mit 1-(2-Hydroxyphenylazo)-2-naphtol (Hyphan) als funktioneller Gruppe und nachfolgende Bestimmung durch Röntgenfluorescenzanalyse", Z. Anal. Chem. **286** (1977) 191.

[59] *P. Burba, B. Gleitsmann,* and *K. H. Lieser,* "Abtrennung und Röntgenfluorescenzanalyse von gelöstem Uran aus natürlichem Wasser mittels chelatbildender Cellulose (am Beispiel natürlicher Wasserproben)", Fresenius' Z. Anal. Chem. **289** (1978) 28.

[60] *P. Burba* and *K. H. Lieser,* "Chromatographie an Celluloseaustauschern – Mikrobestimmung von Fe^{3+}-, UO_2^{2+}- und Cu^{2+}-Ionen in wäßrigen Lösungen mit Trennröhren", Fresenius' Z. Anal. Chem. **291** (1978) 205.

[61] *A. I. Busev, I. P. Kharlamov,* and *G. H. Garibjan,* "Chromatography of complex compounds of metals in non-aqueous solutions. Thin-layer chromatographic separation of metallic 8-quinolinolates from excess 8-quinolinol", Anal. Lett. **5** (1972) 243.

[62] *D. O. Campbell,* "High-speed-high efficiency separation of the transplutonium elements by extraction chromatography", J. Inorg. Nucl. Chem. **35** (1973) 271.

[63] *D. O. Campbell,* "Rapid rare earth metal separation of pressurized ion exchange chromatography", J. Inorg. Nucl. Chem. **35** (1973) 3911.

[64] *R. M. Cassidy,* "A selective method for elemental sulfur analysis by high-speed liquid chromatography", J. Chromatogr. **117** (1976) 71.

[65] *R. M. Cassidy, V. Miketukova,* and *R. W. Frei,* "Ligand-exchange reactions for the fluorescence detection of nanogram amounts of metal ions on thin-layer chromatography plates", Anal. Lett. **5** (1972) 115.

[66] *M. Cernikova* and *B. Konrad,* "Microestimation of zinc in human blood serum", Biochem. Biophys. Acta **71** (1963) 190.

[67] *E. Cerrai* and *G. Ghersini,* "Reversed-phase extraction chromatography in inorganic chemistry", Advan. Chromatogr. **9** (1970) 3.

[68] *E. Cerrai* and *G. Ghersini,* "Reversed-phase chromatography of alkaline earths on paper treated with di-(2--ethylhexyl)orthophosphoric acid in chloride media", J. Chromatogr. **13** (1964) 211; **15** (1964) 236.

*[69] *C. -C. Chang* and *H. -H. Yang,* "Separation of rare earth elements by reversed-phase paper chromatography, the TBP-HNO_3-NH_4SNC system", C. A. **63** (1965) 3 601 g.

[70] *T. T. Chao, M. J. Fishman,* and *J. W. Ball,* "Determination of traces of silver in water by anion exchange and atomic absorption chromatography", Anal. Chim. Acta **47** (1969) 189.

[71] *S. Chikui,* "Electrometric determination in chromatographic development III. Determination of alkaline earth metals during ascending development on paper, and their separation from alkali metal ions", Jap. Analyst **20** (1971) 167 – Anal. Abstr. **23** (1972) 93.

[72] *E. L. Chiotis, G. A. Welford,* and *R. S. Morse,* "Trace determination of nickel by combining paper chromatography with radiometric precipitation", Mikrochim. Acta (1968) 297.

*[73] *N. M. Chistyakov* and *Z. I. Blagoveshchenskaja,* "Determination of copper, cobalt and nickel in water by paper chromatography", C. A. **59** (1963) 7 238 g.

*[74] *D. S. Choe,* "Determination of zirconium and hafnium by reversed-phase chromatography 2. Determination of zirconium and hafnium by using acetophenone as stationary phase", C. A. **82** (1975) 164 400 z.

[75] *J. F. Connolly* and *M. F. Maguire,* "An improved chromatographic method for determining trace elements in food stuffs", Analyst **88** (1963) 125.

[76] *F. Cornett* and *T. W. Gilbert,* "Separation and identification of metal ions by reversed-phase thin-layer chromatography", Anal. Lett. **4** (1971) 69.

[77] *M. Covello* and *G. Ciampa,* "Quantitative analysis of lithium, potassium and caesium chloride by paper chromatography and flame spectrophotometry", J. Chromatogr. **20** (1965) 201.

*[78] *M. Covello, G. Ciampa,* and *F. Manne,* "Separation and spectrophotometric determination of alkali metals in mineral water", C. A. **69** (1968) 5 109 b.

[79] *D. Cozzi, P. G. Desideri,* and *L. Lepri,* "A study of the hydrogen ion concentration gradient on alginic acid thin-layers", J. Chromatogr. **42** (1969) 532.

[80] *E. Cremer* and *E. Seidl,* "Trennung radioaktiver Anionen mit Dünnfilmsorptographie im Bereich unter 10^{-14}g", Chromatographia **3** (1970) 17.

[81] *R. J. Davenport* and *D. C. Johnson,* "Determination of nitrate and nitrite by forced-flow liquid chromatography with electrochemical detection", Anal. Chem. **46** (1974) 1971.

[82] *A. K. De* and *C. R. Bhattacharyya,* "Reversed-phase partition chromatography on paper impregnated with 2-thenoyltrifluoracetone", Separ. Sci. **6** (1971) 621.

[83] *A. K. De* and *B. K. Pal,* "Synthetic inorganic ion-exchangers XV. Thin layer chromatography of metal ions on thorium tungstate: quantitative separation of HgII from several other metal ions", J. Liquid Chromatogr. **2** (1979) 935.

[84] *T. Deguchi*, "Chromatographic separation and detection of mercury, copper, cadmium, nickel, and zinc with dithizone-impregnated silica gel thin-layers", Jap. Anal. **15** (1966) 352.

[85] *T. Deguchi, A. Hisanaga,* and *H. Naga,* "Chromatographic behavior of inorganic anions on a Sephadex G-15 column", J. Chromatogr. **133** (1977) 173.

[86] *T. Deguchi et al.,* "Dissociation of metal dithizonates during gel chromatography", J. Chromatogr. **173** (1979) 271.

[87] *J. F. Dingman et al.,* "Concentration of heavy metals by complexation on dithiocarbonate resins", Anal. Chem. **46** (1974) 774.

*[88] *M. Draignand,* "Application of paper chromatography to the determination of trace elements in silicates", C. A. **72** (1970) 24 100 v.

[89] *T. R. Dulski,* "Determination of acid concentrations in specialty alloy pickling baths by ion chromatography", Anal. Chem. **51** (1979) 1439.

*[90] *T. Dutkiewicz et al.,* "Chromatographic separation and colorimetric determination of mercury, cadmium, cobalt, nickel, and zinc as dithizonates", C. A. **80** (1974) 127 770 h.

[91] *R. Eberhardt, H. Lehner,* and *K. Schlögl,* "Application of high pressure liquid-liquid chromatography in the chemistry of metallocenes", Mh. Chem. **104** (1973) 1409.

[92] *S. Elchuk* and *R. M. Cassidy,* "Separation of the lanthanides on high-efficiency bonded phases and conventional ion-exchange resins", Anal. Chem. **51** (1979) 1434.

[93] *C. T. Enos, G. L. Geoffroy,* and *T. H. Risby,* "High-pressure liquid chromatographic separations of some rhodium and iridium triphenylphosphine complexes", Anal. Chem. **48** (1976) 990.

[94] *C. T. Enos, G. L. Geoffroy,* and *T. H. Risby,* "High-pressure liquid chromatographic separations of some transition metal cluster complexes", J. Chromatogr. Sci. **15** (1977) 83.

[95] *W. J. Evans* and *M. F. Hawthorne,* "High-pressure liquid chromatography of metallocarboranes", J. Chromatogr. **88** (1974) 187.

*[96] *N. E. Fedorova* and *P. N. Kovalenko,* "Determination of small amounts of niobium by partition paper chromatography in ores and minerals", C. A. **72** (1970) 8 960 v.

*[97] *N. E. Fedorova, N. P. Stepanova,* and *V. I. Eiger,* "Paper chromatographic determination of trace amounts of zirconium and hafnium", C. A. **79** (1973) 38 263 g.

[98] *P. Figura* and *B. McDuffle,* "Characterization of the calcium form of Chelex-100 for trace metal studies", Anal. Chem. **49** (1977) 1950.

*[99] *S. Fisel, H. Franchevici,* and *G. Balan,* "Quantitative determination of tin by the chromatographic method on paper and on a cellulose column", Anal. Abstr. **9** (1962) 609.

[99 a] *L. Fishbein,* "Chromatographic and biological aspects of organomercurials", Chromatographic Rev. **13** (1970) 83.

[99 b] *T. Florence* and *G. E. Batley,* "Trace metal species in sea-water I. Removal of trace metals from sea-water by a chelating resin", Talanta **23** (1976) 179.

[99 c] *T. Florence* and *G. E. Batley,* "Determination of the chemical reference to copper, lead, cadmium, and zinc", Talanta **24** (1977) 151.

*[100] *G. Floret* and *V. Massa,* "Determination of cobalt, copper, nickel, and manganese after chromatographic separation of their complexes with 1-(2-pyridylazo)-2-naphthol", C. A. **70** (1969) 31 731 lp.

*[101] *A. A. Fomin,* "Determination of trace elements in biological materials by paper partition chromatography", C. A. **83** (1975) 4 045e.

[102] *M. Fonia* "Rapid polarography in flowing solutions", Ann. Chim. **63** (1973) 763.

[103] *R. Frache* and *A. Dadone,* "Thin-layer chromatography of inorganic ions II. Behavior on ion exchange resins as a stationary phase", Chromatographia **4** (1971) 156.

[104] *M. C. Franks* and *D. L. Pullen,* "Technique for the determination of trace anions by the combination of a potentiometric sensor and liquid chromatography, with

particular reference to the determination of halides", Analyst **99** (1974) 503

[105] *D. J. Freed,* "Flame photometric detector for liquid chromatography", Anal. Chem. **47** (1975) 186.

[106] *R. W. Frei,* "A thin-layer chromatographic method for determination of trace elements in cereal", J. Chromatogr. **34** (1968) 563.

[107] *R. W. Frei* and *V. Miketukova,* "A reflectance spectroscopic study of 4-(2-thiazolyl-azo) resorcinol (TAR) as a spray reagent for cobalt, copper, and nickel on thin-layer chromatograms", Mikrochim. Acta (1971) 290.

[108] *R. W. Frei* and *D. E. Ryan* "Trace metal analysis by combined thin-layer chromatography and reflectance-spectroscopy", Anal. Chim. Acta **37** (1967) 187.

[109] *R. W. Frei* and *H. Stillman,* "A combination of thin-layer chromatography and emission spectroscopy for the microdetermination of alkaline earth metals", Mikrochim. Acta (1970) 184.

[110] *R. W. Frei* and *C. A. Stockton,* "A combination of ring-oven-circular chromatography for trace metal analysis", Mikrochim. Acta (1969) 1196.

[111] *R. W. Frei, R. Liiva,* and *D. E. Ryan* "Reflectance spectroscopic determination of cobalt, nickel and copper with pyridine-2-aldehyde-2-quinolylhydrazone", Can. J. Chem. **46** (1968) 167.

[112] *R. W. Frei, D. E. Ryan,* and *C. A. Stockton,* "The chromatographic properties of transition metal complexes of pyridine-3-aldehyde-2-quinolylhydrazone", Anal. Chim Acta **42** (1968) 59.

[113] *J. S. Fritz,* "Chromatographic separation of metal ions on low capacity, macroreticular resins", Anal. Chem. **46**, (1974) 825.

[114] *J. S. Fritz* and *L. Goodkin,* "Separation and determination of tin by liquid-solid chromatography", Anal. Chem. **46** (1974) 959.

[115] *J. S. Fritz* and *J. Sherma,* "Reversed-phase paper chromatography of metal ions with phenylhydroxamic acid", J. Chromatogr. **25** (1966) 153.

[116] *M. M. Frodyma, D. F. Zaye,* and *T. van Lieu* "Cation analysis by thin-layer chromatography and reflectance spectroscopy Part II The determination of copper, nickel and zinc", Anal. Chim. Acta **40** (1968) 451.

[117] *W. Funasaka, T. Hanai,* and *K. Fujimura,* "High speed liquid chromatographic separations of phthalic esters,carbohydrates, TCA organic acids and organic mercury compounds", J. Chromatogr. Sci **12** (1974) 517.

[118] *E. Gagliardi* and *A. Durst,* "Toluol-3,4-dithiol und verwandte 1, 2-Dithiole als Chelatbildner für Metalle", Monatsh. Chemi **102** (1971) 308.

[119] *E. Gagliardi* and *W. Likussar,* "Trennung der Erdalkalien mittels Dünnschichtchromatographie", Mikrochim. Acta (1965) 765.

[120] *E. Gagliardi* and *W. Likussar,* "Beiträge zur quantitativen Dünnschichtchromatographie V. Die Bestimmung des Wismuts mit MDCM und Abtrennung von den Metallen der Schwefelwasserstoffgruppe", Mikrochim. Acta (1967) 555.

[121] *E. Gagliardi* and *G. Pokorny,* "Beiträge zur quantitativen Dünnschichtchromatographie II. Trennung der Ammoniumsulfidgruppe- MDCM als Reagens – Bestimmung des Nickels"Mikrochim. Acta (1966) 577.

[122] *E. Gagliardi* and *G. Pokorny,* "Beiträge zur quantitativen Dünnschichtchromatographie IV. Spektralphotometrische Bestimmung des Kobalts mit MDCM nach Abtrennung von Metallen der 3. Gruppe", Mikrochim. Acta (1967) 550.

[123] *D. S. Gaibakyan* and *K. S. Babayan* "Use of thin-layer chromatography for identification and separation of rhenium and accompanying elements in alkali solutions", Zavod. Lab. **37** (1971) 9.

[124] *D. S. Gaibakyan* and *R. T. Egikyan,* "Thin-layer chromatography of rare elements IX. Identification and determination of gold III, selenium IV and tellurium IV in an alkaline medium", Ann. Khim. Zh. **23** (1970) 16, C. A. **73** (1970) 31 258 b.

[125] *A. Galik,* "The chromatographic behavior and nature of metal chelates of 1-(2-pyridylazo)-2-naphthol", Anal. Chim. Acta **57** (1971) 399."

[126] *A. Galik,* "Thin layer chromatography of metal chelates Part II. An extended theory and its testing on metal dithizonates and metal diethyldithiocarbamates", Anal. Chim. Acta **67**, (1973) 357.

*[127] *N. Gauchev* and *A. Gaucheva,* "Microchemical analysis on paper, microquantitative determination of Fe", C. A. **63** (1965) 9 050 a.

[128] *E. M. Gel'man,* "Determination of small amounts of selenium and tellurium in mineral materials using paper chromatographic separation", Zh. Anal. Khim. **23** (1968) 736.

*[129] *A. Gereghty,* "Separation of nonvolatile noble metals by reversed-phase extraction chromatography", C. A. **75** (1972) 122 430 j.

[130] *G. R. Gerritse,* "Rapid simultaneous determination of nitrate and nitrite by high-performance liquid chromatography using ultraviolet detection", J. Chromatogr. **171** (1979) 527.

[131] *K. A. Ghosh et al.* "Concentration technique for determintation of air pollutants at sub-micro levels. A new technique for concentration of carbon monoxide", J. Chromatogr. **117** (1976) 29.

[132] *T. W. Gilbert* and *A. Dobbs,* "New detector for ion-exchange chromatography", Anal. Chem. **45** (1973) 1390.

[133] *J. E. Girard,* "Ion chromatography with coulometric detection for the determination of inorganic ions", Anal. Chem. **51** (1979) 836.

[134] *M. Gniazdowski, T. Wasiak,* and *B. Filipowicz,* "Direct colorimetric microdetermination of phosphates in chromatographic spots", Chem. Anal. (Warsaw) **16** (1971) 1367.

[135] *E. J. Going, G. Wesenberg,* and *G. Anderjat,* "Preconcentration of trace metal ions by combined complexation-anion- exchange I. Cobalt, zinc and cadmium with 2-(3'-sulfobenzoyl) pyridine-2-pyridylhydrazone", Anal. Chim. Acta **81** (1976) 349.

*[136] *G. Gorbach,* "Analysis of metal traces by means of microchromatography", C. A. **71** (1969) 45 369j.

[137] *R. E. Graf* and *C. P. Lillya,* "High speed liquid chromatographic analysis of organoiron compounds", J. Organometal. Chem. **47** (1973) 413.

[138] *R. J. T. Graham, L. S. Bark,* and *D. A. Tinsley,* "Quantitative thin-layer chromatography on liquid anion exchangers Part I. An investigation into some of the parameters involved in the direct densitometric determination of zinc", J. Chromatogr. **39** (1969) 211.

[139] *R. J. T. Graham, L. S. Bark,* and *D. A. Tinsley,* "Quantitative thin-layer chromatography on liquid anion exchangers, Part II. A comparison of a spot removal method with an in situ direct densitometric method", J. Chromatrogr. **39** (1969) 218.

[140] *G. Grassini* and *G. Alberti,* "A very sensitive fluorescence test for uranium and its use in the paper-chromatographic estimation of it in natural waters", Microchem. J. Symp. Ser. **2** (1962) 285.

[141] *H. Green,* "Uses of liquid ion-exchangers in inorganic analysis", Talanta **20** (1973) 139.

[142] *J. M. Greenwood et al.,* "Liquid chromatographic separation and quantitative determination of isomeric arene tricarbonyl-chromium complexes", J. Organometal. Chem. **38** (1972) 345.

[143] *Zp. Gregorowicz, J. Kulicka,* and *T. Suwinska,* "Separation and detection of trace amounts of dithizonates of some heavy metals by thin-layer chromatography", Chem. Anal. (Warsaw) **16** (1971) 169.

[144] *M. M. Mota da Guedes, F. G. Römer,* and *B. Griepink,* "Automated separation and preconcentration of copper (II) from natural waters using a column treatment", Z. Anal. Chem. **287** (1977) 19.

[145] *A. C. Handa* and *K. N. Johri,* "Anion determination of sulfide, sulfite, sulfate, and thiosulfate by thin-layer chromatography and ring-colorimetry", Talanta **20** (1973) 219.

[146] *L. D. Hansen et al.,* "Determination of arsenic and sulfur species in environmental samples by ion chromatography", Anal. Chem. **51** (1979) 633.

[147] *N. Hara,* "Simple determination of lead in air using thin-layer chromatographic plates", Ind. Health **11** (1973) 155.

[148] *N. Hara* and *Y. Matsumura,* "Simple ion-exchange thin-layer chromatographic technique for the determination of traces of metals", Ind. Health **9** (1971) 72.

[149] *A. Hartkopf* and *R. Delumyea,* "Use of the luminol reaction for metal ion detection in liquid chromatography", Anal. Lett. **7** (1974) 79.

[150] *M. H. Hashmi* and *A. S. Adil,* "Semiquantitative determination of noble metals by circular thin-layer chromatography", Mikrochim. Acta (1968) 947.

[151] *M. H. Hashmi* and *N. A. Chugtai,* "Semiquantitative determination of anions by circular thin-layer chromatography", Mikrochim. Acta (1968) 1040.

[152] *M. H. Hashmi* and *F. R. Chugtai,* "Semiquantitative determination of thirty cations by circular thin-layer chromatography", Mikrochim. Acta (1971) 924.

[153] *M. H. Hashmi et al.,* "Identification of forty cations and nineteen anions by circular thin-layer chromatography", Anal. Chem. **38** (1966) 1554.

[154] *M. H. Hashmi et al.,* "Semiquantitative determination of micro amounts of iron, copper, cobalt, nickel, and zinc in blood and tissue by circular thin-layer chromatography", Mikrochim. Acta (1968) 712.

[155] *M. H. Hashmi et al.,* "The effect of various factors on micro-analysis by circular thin-layer chromatography", Mikrochim. Acta (1970) 200.

[156] *L. C. Hansen* and *R. E. Sievers,* "Highly permeable open-pore polyurethan columns for liquid chromatography", J. Chromatogr. **99** (1974) 123.

[157] *D. T. Haworth* and *Y. -W. Hung,* "Separation of acetylacetonate chelates by thin-layer chromatography", J. Chromatogr. **108** (1975) 201.

[158] *D. T. Haworth* and *T. Liu,* "High-pressure liquid-chromatographic separation of some organometallic cyclopentadienyl compounds", J. Chromatogr. Sci. **14** (1976) 519.

[159] *D. T. Haworth* and *T. Liu,* "Acetylation of ferrocene. Monitoring a chemical reaction by high pressure liquid chromatography", J. Chem. Educ. **53** (1976) 730.

[160] *J. A. Hayden,* "Determination of uranium in electro-refined plutonium by a combined ion-exchange and X-ray fluorescence technique", Talanta **14** (1967) 721.

[161] *P. Heizmann* and *K. Ballschmiter,* "Chromatography of metal chelates VI. High performance liquid solid chromatography of metal chelates of diacetyl-bis-thibenzydrazone", Z. Anal. Chem. **266** (1973) 206.

[162] *P. Heizmann* and *K. Ballschmiter,* "Chromatography of metal chelates VII. High-performance liquid chromatography of metal 1,2-diketobishiobenzydrazones, metal

dialkyldithiocarbamates and metal 1,2-diketobishthiosemicarbazones", J. Chromatogr. **137** (1977) 153.

*[163] *M. Hejtmanek,* "Quantitative chromatography of inorganic substances I. Principles of photometric evaluation of paper chromatographs", C. A. **61** (1964) 1 246 h.

*[164] *M. Hejtmanek,* "Quantitative chromatography of inorganic substances II. Chromatographic microdetermination of nickel, cobalt, and copper by photometric in situ", C. A. **61** (1964) 1 247 a.

[165] *M. Hejtmanek* and *E. Hozmanova,* "Zur spektralphotometrischen Bestimmung von Cäsium und Rubidium als Pikrate nach papierchromatographischer Trennung", Mikrochim. Acta (1966) 97.

[165a] *U. Hezel,* "Direct quantitative photometry on thin layer chromatograms", Angew. Chem. **85** (1973) 334; Angew. Chem. Int. Ed. Engl. **12** (1973) 298.

[166] *Y. Hirai, N. Yoza,* and *S. Ohashi,* "A spectrophotometric detector for high-performance liquid chromatography of inorganic polyphosphates", J. Liquid Chromatogr. **2** (1979) 677.

[167] *E. Hohmann, M. Rafizadeh,* and *H. Specker,* "Über chromatographische Trennungen der Seltenen Erden", Z. Anal. Chem. **286** (1977) 50.

[168] *H. Holzapfel, L. V. Lau,* and *G. Werner,* "Trennung der Seltenen Erden durch flüssigen Ionenaustausch III. Dünnschichtchromatographische Trennung der Seltenen Erden", J. Chromatogr. **20** (1965) 580.

[169] *T. Honjo et al.,* "Separation and determination of copper (II) and iron (II) as their monothiothenoyltrifluoracetone complexes with pyridine by extraction spectrophotometry and reversed-phase thin-layer chromatography", Anal. Chem. **49** (1977) 2241.

[170] *E. P. Horwitz* and *C. A. A. Bloomquist,* "High speed-high efficiency separation of the transplutonium elements by extraction chromatography", J. Inorg. Nucl. Chem. **35** (1973) 271.

[171] *E. P. Horwitz* and *C. A. A. Bloomquist,* "High speed radiochemical separations by liquid-liquid chromatography using a constant surface porosity support I. Performance of quaternary ammonium chloride-Zipax system for the elution of Cd (II) with hydrochloric acid, II. Separation of metal ions by using tricaprylmethylammonium chloride on Zipax", J. Chromatogr. Sci. **12** (1974) 11, 200.

[172] *P. M. Houpt,* "An element-specific X-rax fluorescence scanner for thin-layer chromatograms", X-Ray Spectrometry **1** (1972) 37.

[173] *M. Hranisavljevic-Jakovlevic et al.,* "Thin-layer chromatography of inorganic ions III. Separation of some noble metal dithiozonates", Mikrochim. Acta (1965) 141.

[174] *J. F. K. Huber* and *A. M. van Urk-Schoen,* "Rapid separation of alkali metals by column ion-exchange chromatography", Anal. Chim. Acta **58** (1972) 395.

[175] *J. F. K. Huber, J. C. Kraak,* and *H. Veening,* "Rapid separation of metal chelates by column liquid-liquid chromatography using ultraviolet detection", Anal. Chem. **44** (1972) 1554.

[176] *K. S. Hui, B. A. Davis,* and *A. A. Boulton,* "The separation by thin-layer chromatography of trace metals as their tetraphenylporphyrin chelates", J. Chromatogr. **115** (1975) 581.

[177] *S. W. Husain* and *F. Eivazi,* "Thin-layer chromatography of 57 metals ions on inorganic ion-exchanger in mixed solvent systems", Chromatographia **8** (1975) 277.

[178] *K. Ishida* and *R. Kuroda,* "Thin-layer chromatography of metal ions using a weakly basic cellulose ion exchanger DEAE", Jap. Anal. **19** (1970) 81.

[179] *K. Ishida, T. Ito,* and *R. Kuroda,* "Thin-layer chromatographic separation of scandium, rare earths, thorium and uranium," Jap. anal. **17** (1968) 1122.

[180] *D. IshII et al.* "Studies on micro high-performance liquid chromatography, V. Design of a microscale liquid chromatography and its application to cation-exchange separation of alkali metals" J. Chromatogr. **157** (1978) 43.

[181] *S. M. Itani,*" Chromatographic separation and spectrophotometric determination of uranium" Nucl. Sci. Abstr. **24** (1970) 38 710.

*[182] *I. M. Ivanova* and *N. B. Zorov,* "Determination of indium impurity in metallic gallium" C. A. **82** (1975) 38 180 q.

[183] *J. R. Jezorek* and *H. Freiser,* "Metal-ion chelation chromatography on silica-immobilized 8-hydroxyquinoline" Anal. Chem. **52** (1979) 366.

[184] *J. R. Jezorek* and *H. Freiser,* "4-(Pyridylazo)resorcinol-based continuous detection system for trace levels of metal ions" Anal. Chem. **51** (1979) 373.

[185] *K. N. Johri* and *K. Bakshi,* "Thin-layer chromatographic separation of transition metal-2-thenoyltrifluoroacetone chelates" Chromatographia **5** (1972) 309.

[186] *K. N. Johri* and *H. C. Mehra*, "Trace metal analysis by combined thin-layer chromatography incorporating fluorescent support and using oven colorimetry" Microchem. J. **15** (1970) 642.

[187] *K. N. Johri* and *H. C. Mehra* "Microdetermination of gold (III), silver (I) and ruthenium (III) by ring colorimetry after separation by thin-layer chromatography" Mikrochim. Acta (1970) 807.

[188] *K. N. Johri* and *H. C. Mehra*, "Inorganic analysis by combined thin-layer chromatography and ring oven technique incorporating fluorescent support" Separ. Sci. **6** (1971) 741.

[189] *K. N. Johri* and *H. C. Mehra*,“ Quercetin as a spray reagent in inorganic thin-layer chromatography and evaluation of the chromatographed species by ring colorimetry” Mikrochim. Acta (1971) 317.

[190] *K. N. Johri* and *H. C. Mehra*, “Thin-layer chromatographic separation of diphenylthiovioluric acid chelates” Chromatographia **8** (1975) 244.

[191] *K. N. Johri* and *H. C. Mehra*, “Thin-layer chromatographic separation of tropolon chelates” Separ. Sci. **11** (1976) 171.

[192] *K. N. Johri, N. K. Kanshik*, and *K. Bakshi*, “Thin-layer chromatographic separation of copper (II), nickel (II) and cobalt (II) as their thiocarbonato-complexes and determination by ring colorimetry” Chromatographia **5** (1972) 326.

[193] *K. N. Johri, H. C. Mehra*, and *N. K. Kaushik*, “Determination of GeIV, SnII, PbII and ZnII, CdII, HgII by ring colorimetry after separation by thin-layer chromatography” Chromatographia **3** (1970) 347.

[194] *K. N. Johri, K. A. Venugopalan*, and *B. S. Arora*, “Glyoxal dithiosemicarbazone as a complexing and chromogenic spray reagent in inorganic TLC” Chromatographia **12** (1979) 476.

[195] *D. R. Jones* and *S. E. Manahan*, “Atomic absorption detector for chromium-organometallic compounds separated by high speed LC” Anal. Lett. **8** (1975) 569.

[196] *D. R. Jones* and *S. E. Manahan*, “Aqueous-phase high-speed liquid chromatographic separation and atomic-absorption detection of aminocarboxylic acid-copper chelates” Anal. Chem. **48** (1976) 502.

[197] *D. V. Jones* and *S. E. Manahan*, “Detection limits for flame spectrophotometric monitoring of high speed liquid chromatographic effluents” Anal. Chem. **48** (1976) 1897.

[198] *G. J. de Jong, W. T. Kok* and *U. A. T. Brinkman*, “The use of aqueous nitrite solutions in reversed-phase chromatography and liquid-liquid extraction” J. Chromatogr. **135** (1977) 249.

*[199] *B. D. Joshi* and *B. M. Patel*, “Chemical separation and spectographic determiantion of trace amounts of rare earth elements in uranium“ C. A. **75** (1971) 104 748 q.

[200] *B. G. Julin, B. G. Vanderborn*, and *J. J. Kirkland*, “Selective flame emission detection of phosphorus and sulfur in high-performance liquid chromatography” J. Chromatogr. **112**, (1975) 443.

[201] *K. Jung* and *H. Specker*, “Dünnschicht-chromatographische Trennung und Nachweis der Seltenen Erden in Erzen, technischen Produkten und Reinelementen” Z. Anal. Chem. **288** (1977) 28.

[202] *N. Kahn* and *J. C. Van Loon*,“ Atomic absorption spectrometry as a chromatography detector for copper-

amino acid complexes in human serum" J. Liquid Chromatogr. *2* (1979) 23.

[203] *G. J. Kamin, J. W. O' Laughlin,* and *C. V. Banks*, "Separation and determination of zirconium in niobium using methylenebis (di-n-hexylphosphine oxide)" J. Chromatogr. **31** (1967) 292.

[204] *T. Kamiura* and *M. Tanaka*, Determination of nitrate in suspended particulate matter by high-performance liquid chromatography with u. v. detection" Anal. Chim. Acta **110** (1979) 117.

[205] *S. Karajannis, H. M Ortner,* and *H. Spitzy*, Column-chromatographic method for removal of the molybdenum matrix in determination of alkali and alkaline-earth metals in molybdenum and its compounds" Talanta **19** (1972) 903.

[206] *P. Kaukare* and *O. Suovaniemi*," A simple method for determination of phosphate from thin-layer plates" J. Chromatogr. **62** (1971) 485.

[207] *K. Kawabuchi* and *R. Kuroda*, "A combined ion-exchange spectrophotometric method for the determination of molybdenum and tungsten in sea water" Anal. Chim. Acta **46** (1969) 23.

*[208] *K. Kawagaki, H. Kadoki,* and *M. Ono*, "Separation and determination of indium by paper chromatography" C. A. **72** (1970) 74 421 w.

[209] *K. Kawabuchi et al.* "Ion exchange concentration on a chelating resin and atomic absorption spectrophotometric determination of heavy metals in geochemical samples" Bunseki Kagaku **25** (1976) 213.

[210] *K. Kawazu,* "Chromatographic separation of metal ions on a macroreticular cation-exchange resin with hydrochloric acid in mixed solvents" J. Chromatogr. **120** (1976) 171.

[211] *K. Kawazu*, "Comparison of efficiency of cation-exchange resins in the chromatographic separation of metal ions with aqueous acetone-hydrochloric acid solutions" J. Chromatogr. **137** (1977) 381.

[212] *K. Kawazu* and *S. J. Fritz*, "Rapid and continuous determination of metal ions by cation-exchange chromatography" J. Chromatogr. **77** (1973) 397.

[213] *K. Kawazu, M. Shibata,* and *H. Kakiyama*, Chromatographic separation of metal ions on a macroreticular cation-exchange resin with hydrochloric acid in aqueous acetone solution" J. Chromatogr. **115** (1975) 543.

[214] *R. Kellner*, "Über die Einsatzmöglichkeit der Kombination Dünnschicht-Chromatographie-Infrarotspektroskopie zur quantitativen Gemischanalyse von Dialkyldithiocarbamaten" Mikrochim. Acta II, (1975) 253.

[215] *T. Kiba*, "Separation of inorganic ions by reversed-phase partition chromatography" Japan Analyst **15** (1966) 991.

[216] *M. Kiboku,* "Thin-layer chromatographic separation of diethyldithocarbamate chelates" Japan. Anal. **17** (1968) 722.

[217] *W. Kielczewski* and *K. Matusiewicz,* "Determination of microgram amounts of magnesium by impregnation method on a filter paper", Chem. Anal. (Warsaw) **13** (1968) 787.

[218] *W. Kielczewski* and *J. Supinski,* "Determination of microgram amounts of cobalt by paper impregnated method", Chem. Anal. (Warsaw) **8** (1963) 59.

[219] *K.-H. König* and *K. Demel,* "Dünnschichtchromatographie anorganischer Ionen an reinem Zirkonhypophosphat", J. Chromatogr. **39** (1969) 101.

[220] *K.-H. König* and *H. Graf,* "Dünnschichtchromatographie anorganischer Kationen an kristallinem Cerium(IV)phosphatsulfat", J. Chromatogr. **67** (1972) 200.

[221] *K. König et al.,* "Chromatographie von Metallchelaten III. Dünnschicht-chromatographische Trennung disubstituierter Dithiocarbamate", Z. Anal. Chem. **259** (1972) 11.

[222] *K.-H. König et al.,* "Zur Chromatographie von Metallchelaten I. Dünnschicht--chromatographie der Metallchelate des 8-Mercaptochinolins mit den Metallen der 1., 2. und 8. Nebengruppe und der 4. Hauptgruppe", Z. Anal. Chem. **297** (1979) 138.

[223] *K.-H. König et al.,* "Zur Chromatographie von Metallchelaten II. Dünnschicht-Chromatographie der Metallchelaten des 1-Hydroxy-2-pyridinthions mit den Metallen der 8. Nebengruppe", Z. Anal. Chem. **297** (1979) 144.

[223a] *K.-H. König et al.,* "Zur Chromatographie von Metallchelaten III. Hochdruckflüssigkeits-Chromatographie der Metallchelate der 8. Nebengruppe mit 8-Mercaptochinolin und 1-Hydroxy-2-pyridinthion", Fresenius Z. Anal. Chem. **297** (1979) 411.

[224] *H. Koizumi, R. D. McLaughlin,* and *T. Hadeishi,* "High gas temperature furnace for species. Determination of organometallic compounds with a high pressure liquid chromatograph and a Zeeman atomic absorption spectrometer", Anal. Chem. **51** (1979) 387.

*[225] *K. Y. Kokk* and *V. B. Aleskovskii,* "Determination of copper in micro amounts of cadmium selenide by paper precipitation chromatography", C. A. **62** (1965) 3 392f.

[226] *K. Konchiyana, N. Yoza,* and *S. Ohashi,* "Gel chromatography of magnesium phosphate complexes. Automatic monitoring by an atomic absorption flow detector", J. Chromatogr. **147** (1978) 271.

[227] *J. Korkisch* and *A. Sorio,* "Determination of seven trace elements in natural waters after separation by solvent extraction and anion-exchange chromatography", Anal. Chim. Acta **79** (1975) 207.

[228] *J. Korkisch* and *A. Sorio,* "The determination of beryllium in geological and industrial materials by atomic-absorption-spectrometry after cation-exchange separation", Anal. Chim. Acta **82** (1976) 311.

[229] *J. Korkisch et al.,* "Chemical analysis of manganese nodules I. Determination of seven main and trace constituents after anion-exchange separation", Anal. Chim. Acta **83** (1976) 83.

*[230] *A. P. Kostikov, S. N. Egorova,* and *T. I. Bulenkov,* "Separation and determination of copper, nickel, and cobalt by thin-layer chromatography combined with reflectance spectrometry", C. A. **82** (1975) 38 218h.

[231] *H. Krainer et al.,* "Alkali and alkaline earth trace analysis of tungsten and tungsten compounds after tungsten matrix separation on DEAE-Sephadex", Talanta **21** (1974) 933.

[232] *H. Kroschwitz, E. Pungor,* and *S. Ferenczi,* "Trennung anorganischer Phosphate an Anionenaustauscher-Fertigplatten", Talanta **19** (1972) 695.

[233] *I. Krzeczkowska et al.,* "2-Hydrazinothiazole derivates for the detection of μg amounts of Cu, Co, Fe(II) and Fe(III) by paper chromatography II. Quantitative determination of Cu^{2+} and CO^{2+} by direct planimetry of spots", C. A. **67** (1967) 96 473p.

[234] *R. Kumar* and *N. Singh,* "Paper chromatography of metal ion complexes with morpholine-4-carbodithioate", Curr. Sci. **47** (1978) 422.

*[235] *V. I. Kurbatova et al.,* "Use of ion-exchange chromatography for the determining trace impurities in heat resistance alloys and ferroalloys", C. A. **74** (1971) 150 696y.

[236] *R. Kuroda, N. Yoshikumi,* and *K. Kawabuchi,* "Use of DEAE-cellulose in inorganic thin-layer chromatography", J. Chromatogr. **47** (1970) 453.

[237] *R. Kuroda et al.,* "Thin-layer chromatographic behaviour of metals on DEAE-cellulose in oxalic acid and mixed oxalic acid-hydrochloric acid media", J. Chromatogr. **139** (1977) 355.

[238] *K. Kurotaki* and *S. Kawamura,* "TTA-containing silica gel thin-layer chromatography of iron, cobalt, zinc, strontium, zirconium, cesium, and cerium", Radiochem. Radioanal. Letters **21** (1975) 129.

*[239] *M. N. Kuz'micheva,* "Separate determination of chromium and nickel in the air by paper chromatography", C. A. **76** (1972) 37 091a.

[240] *J. H. Larochelle* and *D. C. Johnson,* "Chromatographic determination of chromium(VI) with coulometric detection based on the electroatalysis by adsorbed iodine of the reduction at platinum-electrodes in acidic solutions", Anal. Chem. **50** (1978) 230.

[241] *W. Lautenschläger, S. Pahlke,* and *G. Tölg,* "Anodisch oxydierte Aluminiumoberfläche: Ein neues Adsorbens

in der Dünnfilm-Chromatographie", Z. Anal. Chem. **260** (1972) 203.

[242] *M. Lederer,* "Inorganic thin-layer chromatography", Chromatogr. Rev. **9** (1967) 115.

[243] *M. Lederer* and *C. Polcaro,* "Thin-layer chromatography of inorganic anions on aluminium oxide", J. Chromatogr. **84** (1973) 379.

[244] *M. Lederer* and *B. Rinalduzzi,* "The chromatography of metal ions on polyamide thin-layers", J. Chromatogr. **68** (1972) 237.

[245] *C. Lee et al.,* "The use of chelating resion column for preconcentration of trace elements from seawater in their determination by neutron-activation analysis", Talanta **24** (1977) 241.

[246] *H. R. Leene, G. deVries,* and *U. A. Th. Brinkman,* "The use of tri-n-alkylarsine and -arsine oxide in reversed-phase extraction chromatography", J. Chromatogr. **57** (1971) 173.

[247] *H. R. Leene, G. deVries,* and *U. A. Th. Brinkman,* "Reversed-phase extraction chromatography using solutions of nitric acid as eluants", J. Chromatogr. **80** (1973) 221.

[248] *L. Lepri, P. G. Desideri,* and *R. Mascherini,* "Ion-exchange thin-layer chromatography of polyvalent ions on sodium carboxymethylcellulose and Dowex $50\times4(Na^+)$", J. Chromatogr. **70** (1972) 212.

[249] *L. Lepri et al.,* "The behaviour of cation-exchange thin-layers with alkaline and neutral salt solutions as eluents", J. Chromatogr. **47** (1970) 442.

[250] *J. Lehotay et al.,* "Liquid chromatography of metal complexes of N-disubstituted dithiocarbamic acids III. High-performance liquid chromatography of bisdialkyldithiocarbamate complexes of copper(II), cobalt(II), zinc(II), and lead(II)", J. Chromatogr. **172** (1979) 379.

[251] *M. Lesigang* and *F. Hecht,* "Dünnschichtchromatographie von Salzen an Heteropolysäuren", Mikrochim. Acta (1964) 508.

[252] *M. Lesigang-Buchtela,* "Thin-layer chromatography of inorganic ions – a review of the literature", Österr. Chemiker Ztg. **67** (1966) 115.

[253] *M. Lesigang-Buchtela,* "Dünnschichtchromatographie anorganischer Ionen V. über Zusätze zur Kieselgelschichten", Mikrochim. Acta (1969) 1027.

[254] *M. Lesigang-Buchtela* and *K. Buchtela,* "Dünnschichtchromatographie anorganischer Ionen: Quantitative Bestimmung des Urans", Mikrochim. Acta (1967) 670.

[255] *V. T. Lieu* and *C. A. Handy,* "In situ fluorimetric determination of alkaline earth metal ions resolved on paper", Anal. Lett. **7** (1974) 267.

[256] *B. Limoni* and *G. Schmuckler,* "Gel permeation chromatography of transition metal ions using Bio-Gels", J. Chromatogr. **135** (1977) 173.

[257] *O. Liska, G. Guichon,* and *H. Colin,* "Liquid chromatography of metal complexes of N-disubstituted dithiocarbamic acids I. High-performance liquid chromatography of nickel(II) bisdialkyldithiocarbamates", J. Chromatogr. **171** (1979) 145.

[258] *O. Liska et al.,* "Liquid chromatography of metal complexes of N-disubstituted dithiocarbamic acids II. Identification of nickel(II) bisdialkyldithiocarbamate mixed-ligand complexes", J. Chromatogr. **171** (1979) 153.

[259] *O. Liska,* "Liquid chromatography of metal complexes of N-disubstituted dithiocarbamic acids IV. Separation of mixtures of Zn(II), Cu(II), Mn(II), Ni(II), Pb(II), Cr(III), Co(II), Cd(II), Fe(II) diethyldithiocarbamate complexes by high-performance liquid chromatography", J. Chromatogr. **172** (1979) 384.

[260] *J. W. Liu* and *G. E. Janauer,* "Selective separation by reactive ion exchange III. Preconcentration and separation of oxo anions", Anal. Chim. Acta **79** (1975) 219.

[261] *M. Lohmüller, P. Heizmann,* and *K. Ballschmiter,* "Chromatography of metal chelates VIII. High-performance liquid chromatography of metal dithizonates", J. Chromatogr. **137** (1977) 165.

[262] *K. Lorber, K. Mueller,* and *H. Spitzy,* "Column extraction chromatography with dithizone in o-dichlorobenzene. Possibilities for separating metal traces in the nanogram region", Mikrochim. Acta II (1975) 603.

[263] *E. Ludwig, P. Morgenstern,* and *E. Uhlemann,* "Chelate von ß-Dicarbonylverbindungen und ihren Derivaten, Teil XLVII Dünnschicht-chromatographisches Verhalten von Metallchelaten mit Thiobenzoylmethan und Acetylthioacetaniliden", Anal. Chim. Acta **86** (1976) 157.

*[264] *L. McBride, W. Chorney,* and *J. Skok,* "Determination of boron in water, nutrient media, and chlorella cells", C. A. **77** (1972) 149 276 s.

[265] *W. A. MacCrehan* and *R. A. Durst,* "Measurement of organomercury species in biological samples by liquid chromatography with differential pulse electrochemical detection", Anal. Chem. **50** (1978) 2108.

[266] *D. C. Mah, S. Mah,* and *W. M. Tupper,* "Circular paper chromatographic method for determining trace amounts of cobalt, copper, nickel, and zinc in rocks and soils", Can. J. Earth Sci. **2** (1965) 33.

[267] *H. Malissa, R. Kellner,* and *P. Prokopwski,* "Analyse von Metalltetramethylendithiocarbamidatgemischen durch Kombination von Dünnschicht-Chromatographie und Infrarotspektroskopie", Anal. Chim. Acta **63** (1973) 225.

*[268] *E. Maly,* "Determination of the total silicon level in ashed biological material by paper chromatography", C. A. **65** (1966) 12 542 h.

[269] *A. Mangia et al.,* "High-pressure liquid chromatography of dibenzo-18-crown-6 complexes with mercury(II) halides", Anal. Chim. Acta **92** (1977) 111.

[269a] *R. F. C. Mantoura* and *J. P. Riley,* "The use of gel filtration in the study of metal binding by humic acids and related compounds", Anal. Chim. Acta **78** (1975) 193.

[270] *M. Marhol et al.,* "Selective properties and analytical use of an ion-exchange resin based on phenylvinylphosphoric acid", J. Chromatogr. **102** (1974) 89.

*[271] *N. A. Markina,* "Chromatographic separation of trace amounts of copper, nickel, and cobalt in the presence of iron in the air", C. A. **76** (1972) 37 093c.

*[272] *N. A. Markina,* "Determination of copper, cobalt, and nickel directly on a chromatogram using a SF-10 spectrophotometer", C. A. **79** (1973) 121 563a.

*[273] *J. Maslowska* and *R. Soloniewicz*, "The use of thioacetamide for approximate determination of traces of heavy metals by paper chromatography", C. A. **61** (1964) 4 938g.

[274] *Z. Masoomi* and *D. T. Haworth,* "The thin-layer separation of some metal-EDTA complexes on microcrystalline cellulose", J. Chromatogr. **48** (1970) 581.

[275] *V. Massa,* "Determination of mercury by chromatography and photodensitometry of dithizonate. Control of mercuric oxide ointments", Trav. Soc. Pharm. Monpellier **28** (1968) 203.

[276] *V. Massa,* "Photodensitometric determination of lithium in the presence of other alkaline metals following thin-layer chromatography", Trav. Soc. Pharm. Montpellier **29** (1969) 209.

[277] *V. Massa,* "Determination of silver in the presence of other metals by chromatography and photodensitometry of its dithizonate derivative. Application to drug control", Trav. Soc. Pharm. Montpellier **29**, (1969) 221.

[278] *V. Massa,* "Microdetermination of manganese by chromatography and photodensitometry of its complex with 1-(2-pyridylazo)-2-naphtol", Trav. Soc. Pharm. Montpellier **29** (1969) 257.

*[279] *V. Massa,* "Determination of cadmium in the presence of zinc and other metals by dithizonate chromatography and photodensitometry", C. A. **73** (1970) 38 607.

[280] *V. Massa, P. Susplugas,* and *J. Salabert,* "Microdetermination of alkaline earth cations by fluorometry of their thin-layer chromatograms. Application to the analysis of vegetable ash", Trav. Soc. Pharm. Montpellier **34** (1974) 175.

[281] *H. Matsui,* "An automated method for the determination of trace amounts of metal ions by ion-exchange chromatography. Determination of zinc in waters", Anal. Chim Acta **66** (1973) 143.

[282] *A. D. Matthews* and *J. P. Riley,* "The determination of thallium in silicate rocks, marine sediments and seawater", Anal. Chim Acta **48** (1969) 25.

[283] *G. Maura* and *G. Rinaldi,* "Detection of traces of transition metals by peroxy compounds and sorption on a chelating resin", Anal. Chim. Acta **53** (1971) 466.

[284] *A. Mazzucotelli et al.,* "Ion-exchange separation and atomic-absorption determination of fifteen major, minor and trace elements in silicates", Talanta **23** (1976) 879.

[284a] *J. L. Means, D. A. Crevar,* and *J. L. Amster,* "Application of gel filtration chromatography to evaluation of organometallic interactions in natural waters", Limnol. Oceanogr. **22** (1977) 957.

[285] *H. C. Mehra,* "Micro analysis of Ni(II), Co(II), Pd(II) and Ti(IV), Y(III), Zr(IV) by combined thin-layer chromatography and ring colorimetry", Curr. Sci. **46** (1977) 518.

[286] *H. Meier et al.,* "Zur Anwendung der Papierchromatographie für die Anreicherung von Spurenelementen in der Geochemie", Mikrochim. Acta (1969) 852.

[287] *J. Michal et al.,* "Inorganic Chromatographic Analysis", London etc.: van Nostrand Reinhold 1973.

[288] *L. Michel* and *A. Zatka,* "An electrochemical detector with a dropping mercury electrode for high-performance liquid chromatography", Anal. Chim. Acta **105** (1979) 109.

[289] *V. Miketukova* and *R. W. Frei,* "The thin-layer chromatographic properties of trace elements I. Separation of some metals with special reference to lead", J. Chromatogr. **47** (1970) 427.

[290] *V. Miketukova* and *R. W. Frei,* "The thin-layer chromatographic properties of trace metals II. Separation of some cations of commercially precoated sheets", J. Chromatogr. **47** (1970) 435.

[291] *H. Moeckel* and *B. Masloch,* "Liquid-chromatographic determination of elementary sulphur", Z. Anal. Chem. **281** (1976) 379.

*[292] *G. A. Mora,* "Metallochromic indicators in inorganic paper chromatography for the determination of trace elements in foods", C. A. **66** (1967) 27 755z.

[293] *M. Moriyasu* and *Y. Hashimoto,* "Microdetermination of heavy metal chelates by high-performance liquid chromatography", Anal. Lett. A **11** (1978) 593.

[294] *H. Müller* and *R. Rother,* "Dünnschicht-chromatographische Trennung und spektralphotometrische Bestimmung einiger Übergangsmetalle als Monothio-ß-diketonate", Anal. Chim. Acta **66** (1973) 49.

[295] *R. W. Murray* and *R. J. Passarelli,* "Reversed-phase thin-layer and column chromatography with alkyl amine and quaternary ammonium salt stationary phases", Anal. Chem. **39** (1967) 282.

*[296] *A. S. R. Murty,* "Application of thiourea to the determination of selenium by paper chromatography", C. A. **82** (1975) 67 769u.

[297] *G. V. Myasoedova et al.,* "Thin-layer chromatography of the noble metals 9. Use of modified cellulose for the determination of microgram amounts of silver", Zh. Anal. Khim. **29** (1974) 2252.
[298] *Y. Nagai* and *C. Moroguma,* "Separation and simple determination of copper, iron, nickel, and cobalt for thin-layer chromatography with 8-hydroxyquinoline", Jap. Anal. **15** (1966) 46.
[299] *I. I. Nazarenko* and *M. P. Volynets,* "Thin-layer chromatography in inorganic analysis V. Determination of tellurium in mercury-containing materials", Zh. Anal. Khim. **24** (1969) 1209.
[300] *M. P. Neary, P. Seitz,* and *D. M. Hercules,* "A chemiluminescence detector for transition metals separated by ion exchange", Anal. Lett. **7** (1974) 583.
[301] *R. Neirinckx, F. Adams,* and *J. Hoste,* "Determination of impurities in titanium dioxide by neutron activation analysis Part III Determination of vanadium and aluminium in titanium and titania by preseparation", Anal. Chim. Acta **47** (1969) 173.
[302] *R. Neirinckx, F. Adams* and *J. Hoste,* "Determination of impurities in titanium and titanium dioxide by neutron activation analysis Part IV Determination of trace-impurities in titanium dioxide single crystals", Anal. Chim. Acta **48** (1969) 1.
[303] *G. Nickless,* "Chromatography of inorganic ions", Advan. Chromatogr. **7** (1968) 121.
[304] *U. Niederschulte* and *K. Ballschmiter,* "Analytische Anwendung der Di-(4-phenyl)thiosemicarbazone von 1,2-Diketonen", Z. Anal. Chem. **261** (1972) 191.
*[305] *M. K. Nikitin, V. A. Maslov,* and *V. A. Serkova,* "Concentration of gold from solution by partition chromatography", C. A. **79** (1973) 152 516 s.
[305a] *K. Oguma*, "Thin-layer chromatographic separation of scandium, yttrium, the rare earths, thorium and uranium (IV)", Talanta **15** (1968) 860.
[306] *K. Oguma*, "The separation of certain metal ions by solubilization thin-layer chromatography", J. Chromatogr. **42** (1969) 96.
[307] *K. Oguma* and *R. Kuroda,* "Thin-layer chromatographic behaviour of a number of metal ions on DEAE-cellulose in thiocyanic acid-organic solvent mixtures", J. Chromatogr. **52** (1970) 339.
[308] *T. Okubo,* "Determination of rhenium by a combination of reversed-phase-chromatography and ultraviolet spectrophotometry", Jap. Anal. **24** (1975) 600.
[309] *T. Okumura* and *Y. Nishikawa*, "Chromatography of inorganic anions by using aluminium (III)-morin fluorescent complex as a detection reagent", Jap. Anal. **25** (1976) 419.
[310] *J. W. O'Laughlin et al,* "Reversed-phase partition chromatography of the rare earths using methylenebis

(di(2-ethylhexyl)-phosphinc oxide) as the stationary phase", J. Chromatogr. **24** (1966) 376.

[311] *J. O' Laughlin* and *T. P. O' Brien*, "The high-pressure liquid chromatography of metal chelates of diethyl dithiocarbamic acid and dithizone", Anal. Lett. A **11** (1978) 829.

*[312] *F. P. Paderina, K. M. OL' Shanova,* and *N. M. Morozova*, "Concentration of chromium and lead from highly dilute solutions by precipitation chromatography", C. A. **77** (1972) 13 634 z.

*[313] *V. Pagliai* and *F. Pozzi*, "Determination of the metallic impurities in highly eniched uranium solutions by atomic absorption, with the preliminary separation of uranium by partition chromatography", C. A. **78** (1973) 105 660 d.

[314] *R. A. Pasechnova, A. A. Mokhov,* and *B. F. Yudin,* "Determination of tellurium by bismuthiol II in the presence of cadmium using paper peak chromatography", Zh. Anal. Khim. **29** (1974) 2292.

[315] *A. Paster* and *B. Kabacoff*, "The quantitative analysis of free elemental sulphur by thin-layer chromatographic spectrodensitometry", J. Chromatogr. Sci. **14** (1976) 572.

*[316] *K. J. Pawlaczy* and *M. Sierzant,* "Thin-layer chromatography of some metallic contamiantions in pharmaceutical preparations", C. A. **81** (1974) 82473 r.

[317] *I. Pejkovic-Tadic* and *M. Hranisavljevic-Jakovljevic*, Thin-layer chromatography of inorganic ions V. Separation of metal l-nitrose-2-naphtholates", Mikrochim. Acta (1965) 940.

*[318] *V. M. Pensionerova* and *V. E. Pankova*, "Microchemical determination of selenium and tellurium in sulfide minerals using paper partition chromatography", C. A. **69** (1968) 56 791 a.

[319] *R. J. Phillips* and *J. S. Fritz*, "Chromatography of metal ions with a thioglycolate chelating resin", Anal. Chem. **50** (1978) 1504.

*[320] *J. Plamondon,* Rapid determination of uranium in geochemical samples by paper chromatography", C. A. **69** (1968) 15 918 d.

[321] *C. Pohlandt* and *J. S. Fritz*, "Separation of metal ions on a new amide resin", J. Chromatogr. **176** (1979) 189.

[322] *G. Pokórny* and *W. Beyer*, "Beiträge zur quantitativen Dünnschichtchromatographie VII. Spektralphotometrische Bestimmung des Eisens mit MDCM nach Abtrennung von Metallen der Ammoniumsulfidgruppe", Mikrochim. Acta (1968) 582.

*[323] *E. Popper et al.,* "New organic reagent in paper chromatography. Application to inorganic substances. Determination of cadmium, bismuth, and lead", C. A. **65** (1966) 1353 c.

[324] *M. Lederer* and *T. Prosperi,* "A comparison of some ready-for-use thin-layers for the separation of heavy metals with butanol-3 N HCl", J. Chromatogr. **65** (1972) 460.

[325] *D. Puroshottam*, "Chromatographic separation and estimation of traces of uranium in soils and plant ash", J. Sci. Ind. Res. (India) **19 B** (1960) 449.

[326] *M., Qureshi, J. P. Rawat,* and *V. Sharma,* "Chromatographic behaviour of 47 metal ions on titanium (IV) arsenate papers", Talanta **20** (1973) 267.

[327] *M. Qureshi, K. G. Varshney,* and *R. P. S. Rajput,* "Thin-layer chromatography of 49 metal ions on stannic antimonate in aqueous and mixed solvents systems containing dimethyl sulfoxide: qualitative separation of uranium from numerous metal ions", Sep. Sci. **11** (1976) 533.

[328] *J. Rai* and *V. P. Kukreja,* "Thin-layer chromatography of diethyldithiocarbamate complexes of inorganic ions", Chromatographia **3** (1970) 41.

[329] *J. Rai* and *V. P. Kukreja,* "Thin-layer chromatographic separation of gallium, indium, thallium, zinc, and cadmium as their diethyldithiocarbamate complexes", Chromatographia **3** (1970) 499.

[330] *J. Rai* and *V. P. Kukreja,* "Thin-layer chromatographic separation of gold, selenium, tellurium, platinum, and palladium as their diethyldithiocarbamate complexes", Chromatographia **3** (1970) 500.

[331] *A. L. J. Rao* and *C. Sheklar,* "Thin-layer chromatography of inorganic ions as xanthates", Z. Anal. Chem. **277** (1975) 126.

[332] *I. S. Razina* and *M. E. Viktorova,* "Concentration and separation of the platinum group elements and gold using ion-exchange and paper partition chromatography", Zh. Anal. Khim. **25** (1970) 1160.

*[333] *I. S. Razina* and *M. E. Viktorova,* "Determination of submicrogram amounts of osmium and ruthenium after separation by partition paper chromatochromatography", C. A. **80** (1974) 33 582 w.

[334] *I. S. Razina* and *M. E. Viktorova,* "Use of paper partition chromatography for the determination of ruthenium and osmium in mineral raw materials", Zh. Anal. Khim. **29** (1974) 2254.

[335] *R. N. Reeve,* "Determination of inorganic main group anions by high-performance liquid chromatography", J. Chromatogr. **177** (1979) 393.

[336] *K. Reimann,* "Quick identification of heavy metal ions in water and sludge by means of thin-layer chromatography using dithizone", Z. Wasser-Abwasser Forsch. **5** (1972) 3.

*[337] *H. Romanowski* and *D. Radzik*, "Chromatographic separation of zinc with 8-mercaptoquinoline and its complexometric determination in toxicological analysis", C. A. **64** (1966) 18 002 c.

[338] *D. I. Ryabchikov, M. P. Volynets,* and *L. A. Kopneva,* "Thin-layer chromatography in inorganic analysis IV. Microchemical analysis of cosmic dust", Zh. Anal. Khim. **24** (1969) 72.

[339] *K. Saitoh* and *N. Suzuki,* "Gel chromatography of acetylacetone and its metal (II, III) complexes in the Merckogel OR-2000-tetrahydrofuran system", J. Chromatogr. **109** (1975) 333.

[340] *K. Saitoh, M. Satoh,* and *N. Suzuki,* "Distribution coefficients of acetylacetone and tris-(acetylacetonate)-chromium(III) in the Merckogel OR-2000-chloroform system", J. Chromatogr. **92** (1974) 291.

[341] *S. B. Savrin et al.,* "Chelate sorbents for concentration and separation of noble metals", J. Chromatogr. **102** (1974) 287.

[342] *E. Sawicki, J. D. Mulik,* and *E. (Ed.) Wittgenstein,* "Ion chromatographic analysis of environmental pollutants", Ann. Arbor Sci. (1978).

[343] *H. Seiler,* "Anorganische Dünnschicht-Chromatographie 7. Mitt. Quantitative Bestimmung dünnschicht-chromatographisch getrennter Kationen", Helv. Chim. Acta **XLVI** (1963) 2629.

[344] *H. Seiler* and *M. Seiler,*"Anorganische Dünnschicht-Chromatographie 2. Mitt. Nachweis von U(VI) und Ga(III) in Kationen-Gemischen", Helv. Chim. Acta **XLIV** (1961) 939.

[345] *H. Seiler* and *M. Seiler,* "Dünnschichtchromatographie als Hilfsmittel in der Radiochemie 2. Mitt. Untersuchung von Produkten der Neutronenbestrahlung einiger anorganischer Salze", Helv. Chim. Acta **L** (1967) 2477.

[346] *J. G. Sen Gupta,* "Determination of microgram amounts of the six platinum-group metals in iron and meteorites", Anal. Chim. Acta **42** (1968) 481.

[347] *H. J. Senf,* "Applications of thin-layer chromatography in inorganic analysis", Z. Chem. **6** (1966) 102.

[348] *H. J. Senf,* "Dünnschichtchromatographie in der anorganischen Analyse. Die dünnschichtchromatographische Trennung der Diäthyldithiocarbamate der Elemente Quecksilber, Blei, Kupfer, Wismut und Cadmium", J. Chromatogr. **21** (1966) 363.

[349] *H. J. Senf,* "Dünnschichtchromatographie in der anorganischen Analyse 2. Mitt. Dünnschichtchromatographische Trennung einiger 1-Nitroso-2-naphthol-Metallchelate", J. Chromatogr. **27** (1967) 331.

[350] *H. J. Senf,* "Dünnschichtchromatographie in der anorganischen Analyse III Trennung und Nachweis von Arsen, Antimon und Zinn als Pyrrolidindithiocarbamate", Mikrochim. Acta (1968) 954.

[351] *H. J. Senf,* "Dünnschichtchromatographie in der anorganischen Analyse 4. Mitt. Selektiver Nachweis einiger Schwermetalle nach Extraktion als Salicylaldoximkomplexe", Mikrochim. Acta (1969) 522.

[352] *M. D. Seymour* and *J. S. Fritz,* "Determination of metals in mixed hydrochloric and perchloric acids by forced-flow anion exchange chromatography", Anal. Chem. **45** (1973) 1394.

[353] *M. D. Seymour* and *J. S. Fritz,* "Rapid, selective method for lead by forced-flow liquid chromatography", Anal. Chem. **45** (1973) 1632.

[354] *M. D. Seymour et al.,* "Application of forced-flow liquid chromatography to the determination of iron", Anal. Chem. **43** (1971) 1734.

*[355] *R. J. Sheinina, U. K. Khalimova,* and *M. P. Begmatova,* "Determination of residual quantitaties of magnesium chlorate in cotton seeds, their pods and the cotton oil cake", C. A. **85** (1976) 138 116 d.

[356] *J. Sherma,* 'Aqueous-organic solvent mixtures in thin-layer chromatographic separations of metals with ion-exchange resins as adsorbents", Chemist-Analyst **55** (1966) 86.

[357] *M. Shibukawa, K. Oguma,* and *R. Kuroda,* "Thin-layer chromatography of various β-diketones and their metal complexes on styragel 60 A", J. Chromatogr. **166** (1978) 245.

[358] *M. Shibukawa, K. Oguma,* and *R. Kuroda,* "Thin-layer gel chromatography of 8-hydroxyquinoline metal complexes", J. High Resol. Chromatogr. **2** (1979) 229.

[359] *T. Shigematsu et al.,* "Fluorometric determination of cerium in sea-water", Bunseki Kagaku **20** (1971) 575 – Anal. Abstr. **23** (1972) 2862.

[360] *T. Shimizu* and *K. Ikeda,* "The reversed-phase extraction chromatography of the rare earths, thorium, uranium and zirconium with a high-molecular-weight amine in sulphuric acid and ammonium sulphate medium", J. Chromatogr. **85** (1973) 123.

[361] *T. Shimizu* and *R. Ishikura,* "The reversed-phase thin-layer chromatographic behaviour of the rare earths, thorium, uranium, and zirconium with a high-molecular-weight amine in sulfuric acid and ammonium sulphate systems", J. Chromatogr. **56** (1971) 95.

[362] *T. Shimizu, K. Nakazawa,* and *T. Kikuchi,* "The chromatography of a number of metals on cellulose phosphate layers in sulfuric acid and ammonium sulfate media", Chromatographia **9** (1976) 574.

[363] *T. Shimizu, T. Shimojo,* and *A. Nakamura,* "Thin-layer chromatography of inorganic ions an sulphoethyl cellulose in sulphuric acid and ammonium sulphate media", Chromatographia **12** (1979) 135.

[364] *G. J. Shmanenkova, T. J. Romantseva,* and *G. P. Bleshakova,* "Determination of rare earths in cobalt- and copper-based magnetic alloys", Zh. Anal. Khim. **29** (1974) 1545.

[365] *V. Sixta, M. Miksovsky,* and *Z. Sulcek,* "Determination of barium in waters by atomic absorption spectrome-

try after ion-exchange chromatography", Collect. Czech. Chem. Commun. **38** (1973) 3418.

[366] *N. I. Slezko, O. V. Chashchina,* and *A. G. Synkova,* "Chromatopolarographic determination of copper, cadmium, and lead in highly pure bismuth", Zavod. Lab. **41** (1975) 13 – C. A. **83** (1975) 21 549 q.

[367] *H. Small, T. S. Stevens,* and *W. C. Bauman,* "Novel ion-exchange chromatographic method using conductometric detection", Anal. Chem. **47** (1975) 1801.

[368] *F. C. Smith, A. McMurtrie,* and *H. Galbraight,* Ion chromatographic determination of sulfur and chlorine using milligram and submilligram samples", Microchem. J. **22** (1977) 44.

[369] *J. D. Smith,* "Spectrophotometric determination of traces of tin in rocks, sediments and soils", Anal. Chim. Acta **57** (1971) 371.

*[370] *A. Smoczkiewicz* and *W. Mizgalski,* "Determination of zinc in blood serum", C. A. **55** (1961) 24 895 c.

*[371] *A. Smoczkiewicz* and *T. Smoczkiewicz-Szczepanska,* "Microdetermination of copper in pharmaceutical raw materials and preparations", C. A. **57** (1962) 4 763 d.

[372] *V. V. Smolyaninova,* "Electrophoresis and thin-layer chromatography of organic base polythionates", J. Chromatogr. **53** (1970) 337.

[373] *Z. Soljic, S. Turina,* and *V. Marjanovic,* "Separation and detection of Cu, Pb, Bi, Cd, Hg, As, Sb and Sn by thin-layer chromatography on cellulose", Z. Anal. Chem. **258** (1972) 31.

[374] *A. G. Souliotis,* "Simultaneous routine determination of copper and zinc in plants by neutron-activation analysis", Analyst **94** (1969) 359.

[375] *H. Specker, K. Jung,* and *W. Weuster,* "Neue Ergebnisse zur dünnschicht- und säulenchromatographischen Trennung und Anreicherung der Seltenen Erden (Lanthaniden)", G-I-T Fachz. Lab. **23** (1979) 366.

*[376] *V. Spevackova* and *M. Krivanek,* "Dithizone as a stationary phase in reversed-phase chromatography used in activation analysis", C. A. **74** (1971) 49 263 e.

[377] *S. P. Srivastava, V. K. Dua,* and *K. Gupta,* "TLC separation of some inorganic ions on p-toluidine impregnated layers", Z. Anal. Chem. **292** (1978) 415.

[378] *S. P. Srivastava* and *V. K. Gupta,* "Chromatographic separation of some metal ions on NTA-impregnated thin-layer plates", Chromatographia **12** (1979) 496.

[379] *P. Süry,* "Quantitative Mikroanalyse von Kupfer und Aluminium durch Papierchromatographie", Z. Anal. Chem. **250** (1970) 190.

[380] *Z. Sulcek* and *V. Sixta,* "The use of silica gel for the separation of traces of uranium", Anal. Chim. Acta **53** (1971) 235.

[381] *Z. Sulcek* and *V. Sixta,* "Separation of traces of bismuth from excess of iron, antimony, and copper", Collect. Czech. Chem. Commun. **37** (1972) 1993.

[382] *Z. Sulcek, P. Povondra,* and *V. Kratochvil,* "Analytische Schnellmethoden zur Untersuchung von Metallen und anorganischen Rohstoffen XVIII. Bestimmung von Wismutspuren in Galenites und reinem Blei", Collect. Czech. Chem. Commun. **34** (1969) 3711.

[383] *H. Sulser,* "Papier-chromatographischer Nachweis und approximative Bestimmung von Metallspuren in Lebensmittelfarbstoffen", Mitt. Gebiete Lebensm. Hyg. **57** (1966) 66.

[384] *N. Suzuki* and *K. Saitoh,* "Gel chromatography of β-diketones and their metal complexes III. The chromatographic behaviour of metal (II, III) chelates with acetylacetone in systems of a poly(vinyl acetate) gel and various eluting solvents", Bull. Chem. Soc. Japan **50** 2907 (1977).

[385] *N. Suzuki* and *K. Saitoh,* "Gel chromatography of β-diketones and their metal complexes VI. Elution behaviour of various β-diketones and their chromium(III) chelates on different gels", J. Chromatogr. **138** (1977) 79.

*[386] *M. Suzuki, F. Kaiho,* and *S. Takitani,* "Thin-layer chromatography of inorganic compounds XIV. Colorimetric determination of metal dithizonates by elution from the thin-layer plate", C. A. **83** (1975) 187 955 u.

[387] *N. Suzuki, J. Suzuki,* and *K. Saitoh,* "Liquid chromatographic separation of metal(II) chelates of thiothenoyltrifluoroacetone on polystyrene beads", J. Chromatogr. **177** (1979) 166.

[388] *P. Schiller* and *G. B. Cook,* "Determination of trace amounts of gold in natural sweet waters by nondestructive activation analysis after preconcentration", Anal. Chim. Acta **54** (1971) 364.

[389] *E. Schulek et al.,* "Collection of traces of metals on carboxycellulose cation exchangers", Talanta **16** (1969) 323.

[390] *I. Schuphan, K. Ballschmiter,* and *G. Tölg,* "Polychlorierte Verbindungen und ihre Verwendung zur gas-chromatographischen Bestimmung von Elementen im ng-Bereich I. Polychlorierte Xanthogenate und ihre Anwendung zur Nickel-Bestimmung", Z. Anal. Chem. **255** (1971) 116.

[391] *I. Schuphan* and *K. Ballschmiter,* "Polychlorierte Verbindungen und ihre Verwendung zur gas-chromatographischen Bestimmung von Elementen im ng-Bereich II. Polychlorierte Dithiocarbamate", Z. Anal. Chem. **262** (1972) 183.

[392] *G. Schwedt,* "Anorganische Spurenanalyse mit der Hochdruck-Flüssigkeits-Chromatographie am Beispiel des Selens", Fresenius Z. Anal. Chem. **288** (1977) 50.

[393] *G. Schwedt,* "Zur Anwendung der Hochdruck-Flüssigkeits-Chromatographie in der anorganischen Analyse II. Trennung von Metalldiäthyldithiocarbamaten mit der 'reversed-phase'-HPLC", Chromatographia **11** (1978) 145.

[394] *G. Schwedt,* "Zur Anwendung der Hochdruck-Flüssigkeits-Chromatographie in der anorganischen Analyse IV. Bestimmung von Chrom(III)- und Chrom(VI)-Ionen im Abwasser als Dithiocarbamatkomplexe", Z. Anal. Chem. **295** (1979) 382.

[395] *G. Schwedt,* "Zur Anwendung der Hochdruck-Flüssigkeits-Chromatographie in der anorganischen Analyse V. Reversed-phase-Chromatographie von Metalldiäthyl- und -tetramethylendithiocarbamaten", Chromatographia **12** (1979) 289.

[396] *G. Schwedt,* "Hochdruck-Flüssigkeits-Chromatographie für Elementspurenanalysen im Wasser", G-I-T Fachz. Lab. **23** (1979) 540.

[397] *G. Schwedt,* "High-performance liquid chromatography in inorganic analysis", Chromatographia **12** (1979) 613.

[397a] *G. Schwedt,* "Liquid column chromatography with chemical derivatizations after separation", Angew. Chem. **91** (1979) 192, Angew. Chem. Int. Ed. Engl. **18** (1979) 180.

[398] *G. Schwedt* and *Ch. Lippmann,* "Dünnschicht-chromatographischer Schnelltest auf toxische Metalle in Wasser", Dtsch. Lebensm. Rdsch. **70** (1974) 204.

[399] *G. Schwedt* and *A. Schwarz,* "Zur Anwendung der Hochdruck-Flüssigkeits-Chromatographie in der anorganischen Analyse III. Bestimmung von Selen in Trink-, Oberflächen- und Abwasser", J. Chromatogr. **160** (1978) 309.

[399a] *C. Steinberg* and *H. H. Stabel,* "Untersuchungen über gelöste organische Substanzen und ihre Beziehungen zu Spurenelementen", Vom Wasser **51** (1978) 11.

[400] *F. W. E. Strelow,* "Application of ion exchange chromatography to accurate determination of lead, uranium, and thorium in tantalo-molybdates", Anal. Chem. **39** (1967) 1454.

[401] *F. W. E. Strelow, C. J. Liebenberg,* and *F. V. S. Toerien,* "Accurate silicate analysis based on separation by ion-exchange chromatography", Anal. Chim. Acta **47** (1969) 251.

[402] *F. W. E. Strelow, C. J. Liebenberg,* and *A. H. Victor,* "Accurate determination of the major and minor elements in silicate rocks based on separation by cation-exchange chromatography on a single column", Anal. Chem. **46** (1974) 1409.

[403] *T. Takeuchi* and *Y. Tsunoda,* "Thin layer chromatography of dithizone complexes", J. Chem. Soc. Jap. Pure Chem. Sect. **88** (1967) 176.

[404] *T. Takeuchi, Y. Suzuki,* and *Y. Yamazaki,* "Thin-layer chromatography of chromium, manganese, iron, cobalt, nickel, and copper ions on Kieselguhr G as stationary phase", Jap. Anal. **18** (1969) 459.

[405] *S. Takitani, N. Fukuoka,* and *Y. Mitsuzawa,* "Inorganic thin-layer chromatography VIII Determination of metal ions by densitometry", Jap. Anal. **15** (1966) 840.

[406] *S. Takitani et al.,* "Inorganic thin-layer chromatography XII. Thin-layer chromatography of metal oxinates", Jap. Anal. **18** (1969) 626.

*[407] *S. Takitani et al.,* "Thin-layer chromatography of inorganic compounds XIII. Determination of metallic dithizonates by densitometry of thin-layer chromatograms", C. A. **73** (1970) 72 748e.

[408] *J. M. Tanzer, M. I. Krichevsky,* and *B. Chassy,* "Separation of polyphosphates by anion exchange thin-layer chromatography", J. Chromatogr. **38** (1968) 526.

[409] *L. R. Taylor* and *D. C. Johnson,* "Determination of antimony using forced-flow liquid chromatography with a coulometric detector", Anal. Chem. **46** (1974) 262.

[410] *S. N. Tewari* and *N. Bhatt,* "Separation and identification of metal dithizonates by thin-layer chromatography and its application in toxicological analysis", Chromatographia **5** (1972) 624.

[411] *S. N. Tewari* and *N. Bhatt,* "Separation and identification of metal dithizonates by thin-layer chromatography and its application in toxiological analysis", Mikrochim. Acta (1973) 337.

[412] *H. Thielemann,* "Thin-layer chromatographic detection limits (semiquantitative determination) of toxic metal ions (lead, mercury, zinc, copper)", Z. Chem. **15** (1975) 110.

[413] *H. Thielemann,* "Thin-layer chromatographic separation, identification and determination limits as well as the semiquantitative determination of halide ions on activated, commercially available films", Z. Chem. **16** (1970) 283.

[414] *T. Todorova, T. Getcheva,* and *B. Mandova,* "Determination microquantitative de nickel sur du papier chromatographique impregniere", Mikrochim. Acta (1974) 509.

[415] *C. A. Tollinche* and *T. H. Risby,* "Liquid chromatography of metal chelates", J. Chromatogr. Sci. **16** (1978) 448.

[416] *J. C. Touchstone* and *M. F. Dobbins,* "Practice of Thin-layer Chromatography, Wiley, New York (1978).

[417] *J. C. Touchstone,* "Thin-layer chromatographic method for the separation of metal ions separations as PAN complexes", Curr. Sci. India **37** (1968) 695.

[418] *N. Turina,* "Rapid determination of lead in used lubricating oils on thin-layer chromatograms", J. Chromatogr. **93** (1974) 211.

[419] *N. Turina*, "Thin-layer chromatography of tin on a cellulose support using fluorescent chloro-complexes", Chromatographia **9** (1976) 513.

[420] *N. Turina* and *S. Turina*, "Trace analysis of lead in mineral oil by TLC", Chromatographia **10** (1977) 97.

[421] *T. Uchida et al.*, "A simple decomposition and chelating resin separation for the determination of heavy metals in silicates by atomic absorption spectrometry", Anal. Chim. Acta **94** (1977) 275.

[422] *P. C. Uden* and *F. H. Walters*, "The high-pressure liquid chromatographic separation of copper(II) and nickel-(II) schiff base chelates on microparticulate silica", Anal. Chim. Acta **79** (1975) 175.

[423] *P. C. Uden* and *I. E. Bigley*, "High-performance liquid chromatography of metal diethyldithiocarbamates with U. V. and argon plasma emission spectroscopic detection", Anal. Chim. Acta **94** (1977) 29.

[424] *P. C. Uden, D. M. Parees*, and *F. H. Walters*, "Analytical separations of Schiff base chelates by reserve phase HPLC on a 10 micron C-18 bonded silica substrate", Anal. Lett. **8** (1975) 795.

[425] *N. S. Vagina* and *M. P. Volymets*, "Thin-layer chromatography in inorganic analysis II. Separation and determination of rare earths and other impurities during the analysis of lanthanide preparations of various degrees of purity", Zh. Anal. Khim. **23** (1968) 521.

[426] *S. J. Valenty* and *P. E. Behnken*, "Determination of tris(2,2'-bibyridyl)ruthenium(II) derivates by reverse phase paired-ion high-performance liquid chromatography", Anal. Chem. **50** (1978) 834.

[427] *B. M. Vanderborght* and *R. E. van Grieken*, "Enrichment of trace metals in water by adsorption on activated carbon", Anal. Chem. **49** (1977) 311.

[428] *R. E. van Grieken, C. M. Bresseleers*, and *B. M. Vanderborght*, "Chelex-100 ion-exchange filter membranes for preconcentration in X-ray fluorescence analysis of water", Anal. Chem. **49** (1977) 1326.

[429] *T. S. Vasundhara* and *D. B. Parihar*, "Microdetermination of organotin compounds by TLC and spectrophotometry using haematoxylin as reagent", Z. Anal. Chem. **294** (1979) 408.

[430] *H. Veening et al.*, "Liquid chromatography of some arene tricarbonylchromium complexes", Chem. Commun. (1969) 1305.

*[431] *M. D. Vens* and *R. Lauwerys*, "Simultaneous determination of lead and cadmium in blood and urine by coupling ion-exchange resin chromatography and atomic absorption spectrophotometry", C. A. **77** (1972) 70 973h.

[432] *K. A. Venugopalan, B. S. Araora*, and *N. K. Kaushik*, "Microanalytical separation and evaluation of platinum metals", J. High Resol. Chromatogr. **1** (1978) 314.

[433] *M. R. Verma* and *J. Rai,* "The chromatographic separation of some platinum group metals, gold, and mercury", J. Less-Common Metals **15** (1968) 237.

[434] *M. E. Viktorova* and *K. G. Isaeva,* "Separation of nickel, cobalt, copper, and iron by paper chromatography and their quantitative determination in sulphide ores and rocks", Zh. Anal. Khim. **25** (1970) 1140.

[435] *M. P. Volymets et al.,* "Thin-layer chromatography in inorganic analysis VI. Determination of rare earth elements in uranyl nitrate of high purity", Zh. Anal. Khim. **24** (1969) 1477.

[436] *R. A. Webb, D. G. Hallas,* and *H. M. Stevens,* "The determination of iron, manganese, zinc, and copper in plant material by paper chromatography and reflectance densitometry", Analyst **94** (1969) 794.

[437] *H. Wehner, S. Al-Murab,* and *M. Stoeppler,* "Extraction-chromatographic separation of neptunium-239 from fission and activation products in the determination of microgram and submicrogram quantities of uranium", Radiochem. Radioanal. Lett. **13** (1973) 1.

[438] *R. A. Wetzel et al.,* "Determination of trace level ions chromatography with concentrater columns", Anal. Chem. **51** (1979) 1532.

[439] *W. Weuster* and *H. Specker,* "Trennung von Seltenen Erden mittels Hochdruckflüssigkeits-Chromatographie", Fresenius' Z. Anal. Chem. **293** (1978) 301.

[440] *G. L. Wheeler* and *P. F. Lott,* "Rapid determination of trace amounts on selenium IV, nitrite and nitrate by high pressure liquid chromatography using naphthalene-2,3-diamine", Microchem. J. **19** (1974) 390.

[441] *R. B. Willis* and *J. S. Fritz,* "Determination of bismuth by forced-flow liquid chromatography", Talanta **21** (1974) 347.

[442] *A. W. Wolkoff* and *R. H. Larose,* "Separation and determination of low concentration of polythionates by high-speed anion exchange liquid chromatography", Anal. Chem. **47** (1975) 1003.

[443] *A. W. Wolkoff* and *R. H. Larose,* "Effect of chain length on the retention time of polythionates in high-pressure liquid chromatography", J. Chromatogr. Sci. **14** (1976) 353.

[444] *H. Yamaguchi et al.,* "High-performance liquid chromatographic separation of linear and cyclic condensed phosphates", J. Chromatogr. **172** (1979) 131.

*[445] *Y. Yamane, M. Miyazaki,* and *H. Iwase,* "Analysis of metals in water II. Analysis of aluminium, beryllium and chromium by oxine, 2-methyloxine extraction and by thin-layer chromatography", C. A. **69** (1968) 45 939 d.

*[446] *Y. Yamane et al.,* "Analysis of metals in water I. Detection of metals by dithizone extraction and thin-layer chromatography", C. A. **68** (1968) 43 060 r.

*[447] *Y. Yamane et al.*, "Analysis of metals in water III. Total analysis of metals by dithizone, oxine, 2-methyloxine extraction and by thin-layer chromatography", C. A. **75** (1971) 52 645 g.

[448] *Y. Yamamoto et al.*, "Gel chromatography of metal acetylacetonates in organic solvents", Anal. Lett. **6** (1973) 451.

[449] *N. Yoza*, "Gel-chromatography of inorganic compounds", J. Chromatogr. **86** (1973) 325.

[450] *N. Yoza et al.*, "Atomic absorption spectrophotometric analysis of condensed phosphates by a flow detector connected with a gel chromatography column", Anal. Lett. **8** (1975) 641.

[451] *N. Yoza et al.*, "Gel chromatographic investigation of the competitive binding of magnesium and copper(II) ions with polyphosphate ions", J. Chromatogr. **152** (1978) 33.

[452] *B. A. Zabin* and *C. B. Rollins*, "Inorganic ion exchangers for thin-layer chromatography", J. Chromatogr. **14** (1964) 534.

[453] *M. Ziegler, L. Ziegeler*, and *H. Winkler*, "Selektivierung geringer Mengen Kupfer mit Polyäthylenimin-Zellulose", Mikrochim. Acta (1970) 1312.

Gas Chromatography

*[454] *R. G. Ackman* and *R. F. Addison*, "Direct determination of elemental phosphurs by gas-liquid chromatography", C. A. **75** (1972) 29 570 v.

[455] *D. F. Adams et al.*, "Inproved sulfur-reacting microcoulometric cell for gas chromatography", Anal. Chem. **38** (1966) 1094.

[456] *J. B. Addison*, "Studies in gas chromatography-chemical ionisation mass spectrometry of some silicate anions", Analyst **104** (1979) 766.

[457] *R. F. Addison* and *R. G. Ackman*, "Direct determination of elemental phosphorous by gas-liquid chromatography", J. Chromatogr. **47** (1970) 421.

[458] *N. K. Aglinlov et al.*, "Gas-chromatographic and mass-spectrometric analysis of silicon tetrabromide", Zh. Analit. Khim. **30** (1975) 1733.

[459] *M. Akiba, Y. Shimoishi*, and *K. Toei*, "The gas chromatographic determination of selenium in steel with 4-chloro-1,2-diaminobenzene", Analyst **100** (1975) 648.

[460] *M. Akiba, Y. Shimoishi*, and *K. Toei*, "Gas chromatographic determination of selenium in pure elemental arsenic and arsenic(III) oxide with 4-nitro-1,2-diaminobenzene", Analyst **101** (1976) 644.

[461] *D. K. Albert*, "Comparison of electron capture and hydrogen flame detectors for gas chromatographic determination of trace amounts of metal chelates", Anal. Chem. **36** (1964) 2034.

[462] *A. W. Archer*, "A gas-chromatographic method for the determination of increased bromide concentration in blood", Analyst **97** (1972) 428.

[463] *W. A. Aue* and *C. G. Flinn*, "A photometric detector for gas chromatography", J. Chromatogr. **142** (1977) 145.

[464] *W. A. Aue* and *H. H. Hill*, "A hydrogen-rich flame ionization detector sensitive to metals", J. Chromatogr. **74** (1972) 319.

[465] *W. A. Aue* and *H. H. Hill*, "Selective determination of hetero-organics by a dual-channel detector based on flame conductivity and emission", Anal. Chem. **45** (1973) 729.

[466] *K. Bächmann* and *J. Rudolph*, "Gas chromatography of inorganic radioactive compounds", J. Radioanal. Chem. **32** (1976) 243.

[467] *K. Bächmann, K. Buttner*, and *J. Rudolph*, "Verbesserung der Nachweisempfindlichkeit bei gas-chromatographischen Trennungen durch 'radiochemische Verstärkung'", Z. Anal. Chem. **282** (1976) 189.

[468] *K. Bächmann et al.*, "Determination of extremely small traces by enhancement of sensitivity", Mikrochim. Acta I, (1978) 125.

[469] *A. Baiker, H. Geisser*, and *W. Richarz*, "Analysis of CO, CO_2, $COCl_2$, HCl and Cl_2 gas mixtures", J. Chromatogr. **147** (1978) 453.

[470] *K. Ballschmiter*, "Inorganic trace analysis combining thin-layer and gas chromatography: Indirect determination of ruthenium as its thiosemicarbazide chelate", J. Chromatogr. Sci. **8** (1970) 496.

[471] *D. Barlthop* and *P. A. Lewis*, "The determination of oxygen-18 to oxygen-16 ratios in inorganic phosphates by gas-liquid chromatographic-mass spectrometric examination of the tri-n-butyl derivative", Analyt **100** (1975) 862.

[472] *R. S. Barratt*, "Determination of trace amounts of nickel by gas-liquid chromatography", Proc. Soc. Anal. Chem. **9** (1972) 86.

[473] *R. S. Barratt*, "Analytical applications of gas chromatography of metal chelates", Proc. Soc. Anal. Chem. **10** (1973) 167.

[474] *R. S. Barratt et al.*, "The determination of traces of nickel by gas-liquid chromatography", Anal. Chim. Acta **59** (1972) 59.

[475] *E. Bayer, H. P. Muller*, and *R. E. Sievers*, "Hexafluoromonothioacetylacetone – a new ligand for gas chromatographic separation of d^8-metals", Anal. Chem. **43** (1971) 2012.

[476] *J. H. Becker, J. Chevallier*, and *J. Spitz*, "Gaschromatographische Zinnbestimmung in Zircaloy", Z. Anal. Chem. **247** (1969) 301.

[477] *R. Becker et al.*, "Abtrennung und radiochemische Bestimmung geringer Mengen seltener Erden in extraterrestischem Material", Z. Anal. Chem. **274** (1975) 1.

[478] *R. Belcher et al.,* "The gas chromatography, thermal analysis and mass spectrometry of fluorinated lead β-diketonates. Determination of traces of lead by the integrated ion current technique", Anal. Chim. Acta **50** (1970) 423.

[479] *R. Belcher et al.,* "Gas-chromatographic separation of nickel(II), palladium(II), and platinum(II)bis(monothiotrifluoracetylacetonates)", J. Chem. Soc. (1970) 1019.

[480] *R. Belcher et al.,* "Preparative gas chromatography of volatile metal compounds I. Separation of aluminium, chromium and iron β-diketonates", Talanta **17** (1970) 455.

[481] *R. Belcher et al.,* "A gas-chromatographic method for the determination of low concentration of chloride ion", Anal. Chim. Acta **57** (1971) 73.

[482] *R. Belcher et al.,* "Volatile metal chelates of monothioacetylacetone", J. Inorg. Nucl. Chem. **33** (1971) 1851.

[483] *R. Belcher et al.,* "Volatile alkaline earth chelates of fluorinated alkanoylpivaloylmethanes. A thermogravimetric, gas chromatographic and mass spectral study", Anal. Chim. Acta **60** (1972) 109.

[484] *R. Belcher et al.,* "Gas chromatography and thermal analysis of fluorinated bis-monothio-β-diketonates", J. Inorg. Nucl. Chem. **34** (1972) 1017.

[485] *R. Belcher et al.,* "Gas chromatography of divalent transition metal chelates", Anal. Chem. **45** (1973) 1197.

[486] *R. Belcher et al.,* "Volatile transition metal complexes of bis-acetylacetone-ethylenediimine and its fluorinated analogues", J. Inorg. Chem. **35** (1973) 1127.

[487] *R. Belcher, A. Khalique,* and *N. I. Stephen,* "The simultaneous determination of copper and nickel by solvent extraction and gas-liquid chromatography with bis-(acetylpivalylmethane)ethylenediimine", Anal. Chim. Acta **100** (1978) 503.

[488] *R. Belcher et al.,* Thermogravimetric and gas-chromatographic characteristics of some phenylmercury(II) compounds", Chromatographia **9** (1976) 201.

[489] *B. Berck, W. E. Westlake,* and *I. A. Gunther,* "Microdetermination of phosphine by gas-liquid chromatography with microcoulometric, thermionic and flame photometric determination", J. Agric. Food Chem. **18** (1970) 143.

[490] *W. J. Biermann* and *H. Gesser,* "The analysis of metal chelates by gas chromatography", Anal. Chem. **32** (1960) 1525.

[491] *J. R. Birk, C. M. Larsen,* and *R. G. Wilbourn,* "Gas chromatographic determination of sulfide, sulfite, and carbonate in solidified salts", Anal. Chem. **42** (1970) 273.

[492] *M. S. Black* and *R. E. Sievers,* "Environmental analysis problems created by unexpected volatile beryllium compounds in various samples", Anal. Chem. **45** (1973) 1773.

[493] *M. S. Black* and *R. E. Sievers,* "Determination of chromium in human blood serum by gas chromatography with a microwave excited emission detector", Anal. Chem. **48** (1976) 1872.

[494] *R. Bock* and *A. Monerjan,*"Separation and gas chromatographic determination of monovalent thallium as the cyclopentadienyl compound", Z. Anal. Chem. **235** (1968) 317.

[495] *R. Bock* and *H. J. Semmler,* "Separation and determination of fluoride ions with the aid of silicon organic compounds", Z. Anal. Chem. **230** (1967) 161.

[496] *R. Bock* and *S. Strecker,* "Separation and gas-chromatographic determination of traces of fluoride", Z. Anal. Chem. **266** (1973) 110.

[497] *D. H. Bollmann* and *D. M. Mortimore,* "Determination of carbon dioxide, hydrogen sulfide, sulfur dioxide, ethan, and propane using a carbon molecular sieve column", J. Chromatogr. Sci. **10** (1972) 523.

[498] *G. H. Booth* and *W. J. Darby,* "Determination by gas-liquid chromatography of physiological levels of chromium in biological tissues", Anal Chem. **43** (1971) 831.

[499] *G. N. Bortnikov et al.,* "Gas chromatography of metal-organic compounds from the IVB group", Chromatographia **4** (1971) 14.

[500] D. T. Bostick and *Y. Talmi,* "Application of the microwave emission spectrometric system as a silicon-selective detector", J. Chromatogr. Sci. **15** (1977) 164.

[501] *H. K. Bothe* and *J. Leonhardt,* "The role of the argon ionisation detector in the analysis of traces of inorganic gases", J. Chromatogr. **19** (1965) 1.

[502] *A. Brassus,* "Herstellung von Gasgemischen zur Eichung eines Chromatographen mit Hilfe einer dynamischen volumetrischen Methode", Analysis **3** (1975) 576.

[503] *E. Bros, J. Laska,* and *M. Kilarska,* "Application of a helium detector in the determination of CO_2, CO, CH_4, COS, and H_2S by gas chromatography", Chem. Anal. (Warsaw) **19** (1974) 1003, Anal. Abstr. **28** (1975) 5 B 21.

[504] *F. Bruner, F. Ciccioli,* and *R. Rastelli,* "The determination of carbon monoxide in air in the parts per billion range by means of a helium detector", J. Chromatogr. **77** (1973) 125.

[505] *F. Bruner, P. Ciccioli,* and *F. Di Nardo,* "Further developments in the determination of sulfur compounds in air by gas chromatography", Anal. Chem. **47** (1975) 141.

[506] *K. Buchtela, F. Grass,* and *G. Müller,* "Radiogaschromatographie von Metallchelaten", J. Chromatogr. **103** (1975) 141.

[507] *C. A. Burgett,* "Gas chromatographic studies of mixed-ligand complexes of divalent cations", J. Chromatogr. Sci. **11** (1973) 611.

[508] *C. A. Burgett*, "The gas chromatography of selenium as the trimethylsilyl derivative", Anal. Lett. **7** (1974) 799.
[509] *C. A. Burgett*, "Gas chromatography of β-diketonates", Sep. Purif. Methods **5** (1976) 1.
[510] *C. A. Burgett* and *J. S. Fritz*, "Separation and quantitative determination of the yttrium group lanthanides by gas-liquid chromatography", Anal. Chem. **44** (1972) 1738.
[511] *C. A. Burgett* and *J. S. Fritz*, "Column phenomena in the gas-liquid chromatography of metal β-diketonates", J. Chromatogr. **77** (1973) 265.
[512] *C. A. Burgett* and *J. S. Fritz*, "Separation and quantitative determination of the cerium group lanthanides by gas-liquid chromatography", Talanta **20** (1973) 363.
[513] *W. C. Butts*, "Gas chromatography of trimethylsilyl derivatives of common anions", Anal. Lett. **3** (1970) 29.
[514] *W. C. Butts* and *C. V. Banks*, "Solvent extraction and gas chromatography of the rare earth mixed-ligand complexes of hexafluoroacetylacetone and tri-n-buty phosphate", Anal. Chem. **42** (1970) 133.
[515] *W. C. Butts* and *W. T. Rainez*, "Gas chromatography and mass spectrometry of the trimethylsilyl derivatives of inorganic anions", Anal. Chem. **43** (1971) 538.
[516] *R. Bye* and *P. E. Paus*, "Determination of alkylmercury compounds in fish tissue with an atomic absorption spectrometer used as a specific gas-chromatographic detector", Anal. Chim. Acta **107** (1979) 169.
[517] *E. Camera* and *D. Pravisani*, "Gas chromatographic analysis of sulphur in a beed powder", Chim. Ind. **47** (1965) 1210.
[518] *C. J. Cappon* and *J. C. Smith*, "Gas-chromatographic determination of inorganic mercury and organomercurials in biological materials", Anal. Chem. **49** (1977) 365.
[519] *T. Cardwell* and *P. S. McDonagh*, "Gas chromatography of some metal O,O′-dialkyldithiophosphates", Inorg. Nucl. Chem. Lett. **10** (1974) 283.
[520] *T. J. Cardwell, D. J. de Sarro*, and *P. C. Uden*, "Gas chromatography of some volatile metal diethyldithiocarbamates", Anal. Chim. Acta **85** (1976) 415.
[521] *Y. K. Chau, P. T. S. Wong*, and *P. D. Goulden*, "Gas chromatography-atomic absorption method for the determination of dimethyl selenide and dimethyl diselenide", Anal. Chem. **47** (1975) 2279.
[522] *Y. K. Chau et al.*, "Determination of tetra-alkyllead compounds in the atmosphere", J. Chromatogr. Sci. **14** (1976) 162.
[523] *Y. K. Chau et al.*, "Determination of tetra-alkyllead compounds in water, sediment, and fish samples", Anal. Chem. **51** (1979) 186.

[524] *J. B. Clegg,* "A solid electrolyte detector for the determination of trace amounts in the presence of argon and other gases", J. Chromatogr. **52** (1970) 367.
[525] *N. Cockle* and *G. R. Fitch,* "The use of Sephadex in gas chromatography columns", Chem. Ind. (1970) 1966.
[526] *M. L. Conti* and *M. Lesimple,* "Separation of hydrogen isotopes by gas-solid chromatography", J. Chromatogr. **29** (1967) 32.
[527] *W. L. Crider,* "Hydrogen flame emission spectrophotometry in monitoring air for sulfur dioxide and sulfuric acid aerosol", Anal. Chem. **37**, (1965) 1770.
[528] *E. Cropper* and *N. A. Putnam,* "Gas chromatographic determination of fluoride in dental creams", J. Soc. Cosmet. Chem. **21** (1970) 533.
[529] *R. M. Dagnall, T. S. West,* and *P. Whitehead,* "The determination of volatile metal chelates by using a microwave-excited emission detector", Analyst **98** (1973) 647.
[530] *A. G. Datar, P. S. Ramanathan,* and *M. Snkar Das,* "Kaolinite, an absorbent in gas-solid chromatography", J. Chromatogr. **93** (1974) 217.
[531] *A. G. Datar, P. S. Ramanathan,* and *M. S. Das,* "Gas-solid chromatographic studies using beryl as adsorbent", J. Chromatogr. **106** (1975) 428.
[532] *E. H. Daughtrey, A. W. Fitchett,* and *P. Mushak,* "Quantitative measurements of inorganic and methylarsenicals by gas-liquid chromatography", Anal. Chim. Acta **79** (1975) 199.
[533] *J. E. Denison* and *H. Freund,* "Separation and determination of arsenic trichloride and stannic chloride by gas chromatography", Anal. Chem. **37** (1965) 1766.
[534] *T. L. C. De Souza* and *S. P. Bhatia,* "Development of calibration systems for measuring total reduced sulfur and sulfur dioxide in ambient concentrations in the parts of billion range", Anal. Chem. **48** (1976) 2234.
*[535] *G. G. Devyathykh et al.,* "Chromatographic analysis of mixtures of volatile inorganic hydrides", Anal. Abstr. **12** (1966) 5009.
[536] *S. Dilli* and *E. Patsalides,* "Gas chromatography of some oxovanadium(IV) tetradentate ß-ketoamine chelates", J. Chromatogr. **130** (1977) 251.
[537] *T. Doran,* "Supercritical chromatography", Proc. Soc. Anal. Chem. **11** (1974) 117.
[538] *H. W. Dürbeck* and *R. Niehaus,* "Gas-chromatographische Simultanbestimmung von O_2, N_2, CO, CO_2, N_2O, SO_2, CH_4, C_2H_4, C_2H_6 im ppm-Bereich I", Chromatographia **11** (1978) 14.
[539] *K. J. Eisentraut* and *R. E. Sievers,* "Volatile rare earth chelates", J. Am. Chem. Soc. **87** (1965) 5254.
[540] *K. J. Eisentraut, J. Griest,* and *R. E. Sievers,* "Ultratrace analysis for beryllium in terrestrial meteoritic, and

Apollo 11 and 12 lunar samples using electron capture gas chromatography", Anal. Chem. **43** (1971) 2003.

[541] *R. Ellerker, F. Lax,* and *H. J. Dee,* "Gas chromatographic separation of anaerobic gases using porous polymers", Water Res. **1** (1967) 243.

[542] *O. Ellren, I. B. Peetre,* and *B. E. F. Smith,* "Gas chromatographic investigation of organometallic compounds and their analogues I. Determination, calculation, and correlation of Kovats' retention indices for tetraalkoxysilanes", J. Chromatogr. **88** (1974) 295.

[543] *C. S. Evans* and *C. M. Johnson,* "The separation of some alkyl selenium compounds by gas chromatography", J. Chromatogr. **21** (1966) 202.

[544] *B. Flegler et al.,* "Verteilungsverhalten einiger Diethyldithiophosphinatochelate: Anwendung zur gaschromatographischen Bestimmung des Cadmiums", Talanta **26** (1979) 761.

[545] *C. G. Flinn* and *W. A. Aue,* "Photometric detection of selenium compounds for gas chromatography", J. Chromatogr. **153** (1978) 49.

[546] *T. M. Florence et al.,* "Beryllium content of NBS standard reference orchard leaves", Anal. Chem. **46** (1974) 1874.

[547] *R. Fontaine et al.,* "Preparation and gas chromatographic analysis of thorium and uranium chelates", Chromatographia **3** (1970) 532.

[548] *R. Fontaine et al.,* "Utilisation des chelates de dicetones-ß par l'analyse quantitative de l'uranium et du thorium", Anal. Chim. Acta **62** (1972) 337.

[549] *R. Fontaine et al.,* "Retention von Metallchelaten in der Gas-Chromatographie. Die relative Bedeutung der Auflösungs- und Adsorptionsphänomene", J. Chromatogr. **104** (1975) 1.

[550] *J. K. Foreman, T. A. Gough,* and *E. A. Walker,* "The determination of traces of beryllium in human and rat urine samples by gas chromatography", Analyst **95** (1970 797.

[551] *G. M. Frame et al.,* "Trace determination of beryllium oxide in biological samples by electron-capture gas chromatography", Anal. Chem. **46** (1974) 534.

[552] *J. A. Fresen, F. H. Cox,* and *M. J. Witter,* "Determination of fluoride in biological materials by means of gas chromatography", Phar. Weekblad Ned. **103** (1968) 909.

[553] *J. S. Fritz* and *R. C. Chang,* "Macroporous resins for sorption and chromatographic separation of gases", Anal. Chem. **46** (1974) 938.

[554] *T. Fujinaga, T. Kuwamoto,* and *T. Kimoto,* "Rapid separation and determination of rare-earth elements by use of the ligand-vapour gas-chromatographic method", Talanta **23** (1976) 753.

[555] *T. Fujinaga, T. Kuwamoto,* and *S. Murai,* "Gas chromatography of metal chelates with carrier gas containing ligand vapour", Talanta **18** (1971) 429.

[556] *T. Fujinaga, T. Kuwamoto,* and *S. Murai,* "Gas chromatography of metal chelates with carrier gas containing ligand vapour. Thorium(IV) trifluoroacetylacetonate", Anal. Chim. Acta **71** (1974) 141.

[557] *G. Garzo, J. Fekete,* and *M. Blasco,* "Determination of the gas chromatographic retention indices of various organometallic compounds", Acta chim. Acad. Sci. Hung. **51** (1967) 359.

[558] *J. Gawlowski, J. Niedzielski,* and *A. Wieckowski,* "Argon ionisation detector sensitive to hydrogen", J. Chromatogr. **151** (1978) 370.

[559] *H. Geissler* and *H. Kriegsmann,* "Separation solvents for the gas chromatographic investigation of certain butyltin compounds", Z. Chem. **5** (1965) 423.

[560] *V. Gemmer-Colos* and *R. Neeb,* "Dialkyldithiocarbamate als Reagentien zur gas-chromatographischen Bestimmung von Metallen II. Bestimmung von Zink, Kupfer und Nickel als Dipropyldithiocarbamatochelate nach Ausschütteln aus wässrigen Lösungen", Z. Anal. Chem. **293** (1978) 290.

[561] *C. Genty et al.,* "Determination of trace quantities of aluminium and chromium in uranium by gas phase chromatography", Anal. Chem. **43** (1971) 235.

[562] *J. N. Gerber* and *D. T. Sawyer,* "Nature of nonspecific interactions in gas-solid chromatography", Anal. Chem. **44** (1972) 1199.

[563] *D. J. Glover* and *J. C. Hoffsommer,* "Gas chromatographic analysis of nitrate and nitrite ions in microgram quantities by conversion to nitrobenzene", J. Chromatogr. **94** (1974) 334.

[564] *A. R. Goldbaum, T. J. Domanski,* and *E. L. Schloegel,* "The use of a helium ionization detector for the determination of atmospheric gases", J. Gas Chromatogr. **6** (1968) 394.

[565] *T. A. Gosink,* "Rapid simultaneous determination of picogram quantities of aluminium and chromium from water by gas phase chromatography", Anal. Chem. **47** (1975) 165.

[566] *D. G. Greer* and *T. J. Bydalek,* "Response characterization of the Melpar flame photometric detector for hydrogen sulfide and sulfur dioxide", Environ. Sci. Technol. **7** (1973) 153.

[567] *H. W. Grice, M. L. Yates,* and *D. J. David,* "Response characteristics of the Melpar flame photometric detector", J. Chromatogr. Sci. **8** (1970) 90.

[568] *W. E. L. Grossman, J. Eng,* and *Y. C. Tong,* "Selectivity of a microwave discharge detector in the gas-chromatographic determination of dimethylmercury", Anal. Chim. Acta **60** (1972) 447.

[569] *S. Grys,* "The gas-liquid chromatographic determination of inorganic iodine, iodide, and tightly bound iodine in milk", J. Chromatogr. **100** (1974) 43.

[570] *B. J. Gudzinowicz* and *J. L. Driscoll,* "Separation of alkyl/aryl and perfluorinated organoarsenic compounds by gas-liquid chromatography", J. Gas Chromatogr. **1** (5), (1963) 25.

[570a] *B. J. Gudzinowicz* and *H. F. Martin,* "Separation of organo- and organobromoarsenic compounds by gas liquid chromatography", Anal. Chem. **34** (1962) 648.

[571] *M. C. Gupta et al.,* "Response of the flame ionization detector to nitrogen and oxygen", J. Chromatogr. Sci. **11** (1973) 373.

[572] *A. G. Hamlin, G. Iveson,* and *T. R. Phillips,* "Analysis of volatile inorganic fluorides by gas liquid chromatography", Anal. Chem. **35** (1963) 2037.

[573] *P. R. Handy* and *A. MacDonald,* "New instrumental combustion method for the microdetermination of phosphorus", J. Assoc. Off. Anal. Chem. **53** (1970) 780.

[574] *L. C. Hansen et al.,* "Rapid analysis for subnanogram amounts of chromium in blood and plasma using electron capture gas chromatography", Anal. Chem. **43** (1971) 349.

[575] *C. H. Hartmann* and *K. P. Dimick,* "Helium detector for permanent gases", J. Gas Chromatogr. **4** (1965) 163.

[576] *T. Hashizume* and *Y. Saski,* "Gas chromatography of sugar phosphates", Anal. Biochem. **15** (1966) 199.

[577] *T. Hashizume* and *Y. Saski,* "Determination of orthophosphate by gas chromatography", Anal. Biochem. **21** (1967) 316.

[578] *T. Hashizume* and *Y. Saski,* "Base and phosphorus composition analysis by gas chromatography", Anal. Biochem. **24** (1968) 232.

[579] *S. Hasinski,* "Gas-chromatographische Spurenbestimmung der Schwefel- und Phosphorverbindungen. Neue Konstruktion des nichtlöschenden Flammenphotometer-Detektors FPD", Chem. Anal. (Warsaw) **20** (1973) 1135.

[580] *R. A. Hasty,* "A gas chromatographic method for the microdetermination of iodine", Mikrochim. Acta (1971) 348.

[581] *R. A. Hasty,* "Gas chromatographic microdetermination of iodine by derivatization with ketones", Mikrochim. Acta (1973) 621.

[582] *F. F. Henry* and *T. M. Thorpe,* "Gas chromatography of the TMS derivatives of arsenic, arsenious, and dimethylarsinic acids", J. Chromatogr. **166** (1978) 577.

[583] *G. W. Heylman* and *J. E. Pikula,* "Analysis of methylfluorosilanes from methylpolysiloxanes by gas chromatography", J. Gas Chromatogr. **3** (1965) 266.

[584] *D. C. Hildebrand* and *E. E. Pickett,* "Determination of the volatility of metal chelates by atomic absorption", Anal. Chem. **47** (1975) 424.

[585] *F. D. Hileman et al.,* "In situ preparation and evaluation of open pore polyurethane chromatographic columns", Anal. Chem. **45** (1973) 1126.

[586] *H. H. Hill* and *W. A. Aue,* "Selective detection of organometallics in gas chromatographic effluents by flame photometry", J. Chromatogr. **74** (1972) 311.

[587] *H. H. Hill* and *W. A. Aue,* "A silicon detector for gas chromatography", J. Chromatogr. **140** (1977) 1.

[588] *R. D. Hill* and *H. Gesser,* "Flame photometric detection of metal chelates separated by gas chromatography", J. Gas Chromatogr. **1** (12), (1963) 14.

[589] *D. W. Hill* and *H. A. Newell,* "The variation with polarising voltage of the response to methane, carbon dioxide and nitrous oxide, of a macro-argon ionisation detector for gas chromatography", J. Chromatogr. **32** (1968) 737.

[590] *R. P. Hirschmann* and *T. L. Mariani,* "Synthetic diamond – a solid adsorbent for corrosive gases", J. Chromatogr. **34** (1968) 78.

[591] *T. Hobo et al.,* "Foam separation of parts per billion of methylmercury chloride for subsequent gas-chromatographic determination", Jap. Analyst **24** (1975) 288.

[592] *R. L. Hoffmann, G. R. List,* and *C. D. Evans,* "Gas-solid chromatographic separation of atmospheric gases on activated aluminia", Nature **211** (1966) 965.

[593] *R. V. Holland* and *P. W. Board,* "Determination of ozone by gas chromatography", Analyst **101** (1976) 887.

[594] *O. L. Hollis,* "Separation of gaseous mixtures using porous polyaromatic polymer beads", Anal. Chem. **38** (1966) 309.

[595] *O. L. Hollis,* "Porous polymers used in GC and LC", J. Chromatogr. Sci. **11** (1973) 335.

[596] *B. Iatridis* and *G. Parissakis,* "Chlorination and gas chromatographic determination of arsenic in oxides, sulphides, ores, and alloys", J. Chromatogr. **122** (1976) 505.

[597] *B. Iatridis* and *G. Parissakis,* "Determination of germanium in oxides, ores, and alloys by chlorination and gas chromatography", Anal. Chem. **49** (1977) 909.

[598] *B. Iatridis, J. Kontoyannakos,* and *G. Parissakis,* "Gas chromatographic determination of tin in oxides, sulphides, minerals, and alloys after chlorination", J. Chromatogr. **178** (1979) 277.

[599] *K. G. Ikels* and *J. R. Neville,* "An improved multigas analyzer", J. Gas Chromatogr. **6** (1968) 222.

[599a] *N. Ives* and *L. Guiffrida,* "Investigation of thermionic response for the gas chromatography of organic compounds containing phosphorus, nitrogen, arsenic, or chlorine", J. Assoc. Offic. Anal. Chem. **50** (1967) 1.

[600] *P. Jacquelot* and *G. Thomas,* "Gas chromatography of vanadyl trifluoroacetylacetonate", Bull. Soc. Chim. France (1970) 3167.

[601] *P. Jacquelot* and *G. Thomas,* "Gas chromatography of mixed complexes of nickel(II) or cobalt(II) with tri-

fluoroacetylacetone and dimethylformamide", Bull. Soc. Chim. France (1971) 702.

[602] *P. Jacquelot* and *G. Thomas,* "Chromatographie en phase gazeuse de quelques chelates de la thenoyltrifluoroacetone, J. Chromatogr. **66** (1972) 121.

[603] *P. Jacquelot, J. P. Meille,* and *G. Thomas,* "Extraction liquide-liquide et chromatographie en phase gazeuse de chelates mixtes de nickel(II) et cobalt(II): complexes metalthenoyltrifluoracetone-diethylamine", Anal. chim. Acta **60** (1972) 335.

[604] *P. G. Jeffery* and *P. J. Kipping,* "The determination of constituents of rocks and minerals by gas chromatography Part I. The determination of carbon dioxide", Analyst **87** (1962) 379.

[604a] *D. Jentzsch* and *E. Otte,* "Detektoren in der Gas-Chromatographie", Akad. Verlagsges., Frankfurt (1970).

[605] *P. Jones* and *G. Nickless,* "The estimation of inorganic mercury at low concentration by GC", J. Chromatogr. **76** (1973) 285.

[606] *P. Jones* and *G. Nickless,* "Determination of inorganic mercury by gas-liquid chromatography", J. Chromatogr. **89** (1974) 201.

[607] *K. C. Joshi* and *V. N. Pathak,* "Metal chelates of fluorinated 1,3-diketones and related compounds", Coord. Chem. Rev. **22** (1977) 37.

[608] *R. S. Juvet* and *R. P. Dubrin,* "Characterization of flame photometric detector for gas chromatography", Anal. Chem. **38** (1966) 565.

[609] *R. D. Kadeg* and *G. D. Christian,* "Gas chromatographic determination of selected group IV-VI element hydrides", Anal. Chim. Acta **88** (1977) 117.

[610] *G. Kaiser et al.,* "Contribution to the optimization of the chelate gas chromatographic determination of beryllium in limited amounts of organic materials", Z. Anal Chem. **259** (1972) 257.

[611] *R. Kaiser,* "Water determination down to 10^{-9} weight percent", Chromatographia **2** (1969) 453.

[611a] *R. Kaiser,* "Reversions-Gas-Chromatographie, ein neues Verfahren der Mikrospurenanalyse flüchtiger Stoffe", Naturwiss. **57** (1970) 295.

[611b] *R. Kaiser,* "Ultra trace analysis of volatile substances with GC methods in the sub ppm and ppb range down to 10^{-10}%", J. Chromatogr. Sci. **9** (1971) 227.

[612] *N. M. Karayannis* and *A. H. Corwin,* "Hyper-pressure gas chromatography VI. Gas chromatography of etioporphyrin-metal(II) chelates", J. Chromatogr. **47** (1970) 247.

[613] *N. M. Karayannis* and *A. H. Corwin,* "Volatilization and separations of metal acetylacetonates at 115 °C by hyperpressure gas chromatography", J. Chromatogr. Sci. **8** (1970) 251.

[614] *N. M. Karayannis et al.*, "Apparatus and materials for hyperpressure gas-chromatography of nonvolatile compounds", Anal. Chem. **40** (1968) 1736.
[615] *H. Kawaguchi, T. Sakamoto,* and *A. Mizuike,* "Emission spectrometric detection of metal chelates separated by gas chromatography", Talanta **20** (1973) 321.
[616] *H. Kawaguchi et al.,* "Gas chromatography of metal chelates of trifluoroacetylacetone using an emission spectrometric detector", Japan Analyst **22** (1973) 1434.
[617] *J. S. Keller, H. Veening,* and *B. R. Willeford,* "Gas chromatographic separation and determination of isomeric methylbenzenetricarbonylchromium complexes", Anal. Chem. **43** (1971) 1516.
[618] *M. Kilarska,* "Use of the electron capture detector EC-1 for chromatographic analysis of volatile sulfur compounds", Chem. Anal. (Warsaw) **15** (1970) 953.
[619] *A. V. Kiselev* and *Y. I. Yashin,* "Gas-adsorption chromatography", New York-London: Plenum Press (1969).
*[620] *A. Kito* and *Y. Miyake,* "Gas chromatography of ceriumgroup lanthanide chelates of dipivaloylmethane", C. A. **82** (1975) 79 965t.
[621] *A. Kito, M. Nakane,* and *Y. Miyake,* "Gaschromatographic and thermogravimetric analyses of metal chelates of 2-picolyl alkyl ketones", Japan Analyst **26** (1977) 19.
[622] *A. Kito, M. Nakane,* and *Y. Miyake,* "Thermogravimetric analyses and gas chromatography of the metal chelates of ß-diketones containing alicyclic side chains I., II.", Japan Analyst **27** (1978) 1.
[623] *A. Kito et al.,* "Gas chromatography of some metal chelates of 3-alkylacetylacetones", Japan Analyst **20** (1971) 1363.
[624] *A. Kleinmann* and *R. Neeb,* "Gas-chromatographische Trennung und Bestimmung einiger Kationen als Diethyldithiophosphinatochelate", Z. Anal. Chem **285** (1977) 107.
[625] *J. Koch* and *K. Figge,* "Zur Analytik von Methylzinnstabilisatoren", J. Chromatogr. **109** (1975) 89.
[626] *H. W. Kohlschütter* and *W. Höppe,* "Gas adsorption chromatography with active chromium(III) oxide", Z. Anal. Chem. **197** (1963) 133.
*[627] *V. A. Komarov*, "β-Diketone und ihre Anwendung in der analytischen Chemie. Gas-chromatographische Bestimmung von Metallchelaten", Z. Anal. Chem. **285** (1977) 137.
[628] *B. R. Kowalski, T. L. Isenhour,* and *R. E. Sievers,* " Ultratrace mass spectrometric metal analysis using heptafluorodimethyloctanedione chelates", Anal. Chem. **41** (1969) 998.
[629] *J. Krupcik,* et al. "Gas chromatographic separation of some dialkyldithiocarbamates of nickel (II)", J. Chromatogr. **112** (1975) 189.

[630] *J. Krupcik* et al. "Analysis of divalent metal dialkyldithiocarbamates by gas chromatography and mass spectrometry", J. Chromatogr. *171* (1979) 285.

[631] *R. Kruse* and *H. A. Rüssel,* "Gas-chromatographische Bestimmung des Arsens als Trimethylsilylarsenat in biologischem Material", Z. Anal. Chem. **286** (1977) 226.

[632] *C. Kutal* and *R. E. Sievers*, "Rate and equilibrium study of the cis-trans isomerisation of tris(1,1,1-trifluoro-2,4-pentanedionato)chromium III in the gas phase", Inorg. Chem. **13** (1974) 897.

[633] *T. Kuwamoto*, "Analytical application of metal chelate compounds in gas chromatography", Japan Analyst **21** (1972) 445.

[634] *T. Kuwamoto*, "Gas chromatography of metal compounds in micro- and submicrogram amounts", C. A. **84** (1976) 83 613k.

[635] *M. L. Lee* and *D. Burrell*, "Soluble aluminium in marine and fresh water by gas-liquid chromatography", Anal. Chim. Acta **66** (1973) 245.

[636] *I. Lindgren* and *B. Jansson,* "Support material effects in gas chromatography analyses of organophosphorus compounds", J. Chromatogr. **106** (1975) 385.

[637] *J. R. Lovett* and *G. F. Lee*, "Analysis of chromium in natural waters by gas chromatography", Environ. Sci. Technol. **10** (1976) 67.

[638] *V. Luckow* and *H. A. Rüssel*, "Gas chromatographic determination of trace amounts of inorganic mercury", J. Chromatogr. **150** (1978) 187.

[639] *K. L. McDonald*, "Anion analysis by gas chromatography", Anal. Chem. **44** (1972) 1298.

[640] *J. MacGree* and *K. G. Allen*, "Alkylation and gas chromatography of aqueous inorganic halides", Anal. Chem. **42** (1970) 1672.

*[641] *V. G. Makarenko*, et al. "Gas-chromatographische Trennung von organometallischen Verbindungen", Z. Anal. Chem. **281** (1976) 164.

[642] *G. Manara* and *M. Taramasso*, "Sieving effects of hornites in gas-adsorption chromatography", J. Chromatogr. **65** (1972) 349.

[643] *G. Manara* and *M. Taramasso*, "Adsorption properties of the surfaces of attapulgite and sepiolite", J. Chromatogr. **77** (1973) 109.

[644] *J. Masaryk* et al. "Gas chromatographic separation of dithiocarbamates of some metals", J. Chromatogr. *115* (1975) 256.

[645] *D. R. Matthews, W. D. Shults,* and *M. R. Guerin*, "Extraction-derivatization gas-chromatographic determination of trace phosphate in aqueous media", Anal. Chem. **43** (1971) 1582.

[646] *A. Meyer* et al. "Extrem nachweisstarkes gaschromatographisches Bestimmungsverfahren für Selen nach Abtrennung durch Verdampfen im Sauerstoffstrom I.

Bestimmung im Reinstkupfer", Z. Anal. Chem. **281** (1976) 201.

[647] *G. Michael, U. Danne,* and *G. Fischer*, "Gaschromatographische Untersuchungen der Siliciumtetrahalogenide", J. Chromatogr. **118** (1976) 104.

[648] *J. W. Mitchell* and *C. V. Banks*, "Volatile complexes of some lanthanides and related elements with fluorinated β-diketones and organophosphorus adducts", Anal Chim. Acta **57** (1971) 415.

[649] *M. Miyazaki* and *H. Kaneko*, "Feasibility of gas-chromatography for ultramicro determination of aluminium in biological materials", Chem. Pharm. Bull. **18** (1970) 1933.

[650] *M. Miyazaki* et al. "Gas chromatography of copper(II) and nickel(II) chelates of some derivatives of acetylacetone and salicalaldehyde", Chem. Pharm. Bull. **14** (1966) 117.

[651] *P. Mushak* et al. "Perhalobenzenesulfinates as reagents in the determination of inorganic mercury in various media by gas-liquid chromatography", J. Chromatogr. **87** (1973) 215.

[652] *A. L. Myerson* and *J. J. Chludzinski*, "A gas chromatographic determination of HCN in combustion effluents in the low ppm range", J. Chromatogr. Sci. **13** (1975) 554.

[653] *S. Nakashima* and *K. Toei*, "Determination of ultramicro amounts of selenium by gas chromatography", Talanta **15** (1968) 1475.

[654] *G. Nota* and *R. Palombari*, "Determination of cyanides and thiocyanates in water by gas chromatography", J. Chromatogr. **84** (1973) 37.

[655] *G. Nota*, et al. "Determination of complex cyanides in water by gas-chromatography", J. Chromatogr. **123** (1976) 411.

[656] *M. H. Noweir* and *J. Cholak*," Gas chromatographic determination of beryllium in biological materials and air", Environ. Sci. Technol. **3** (1969) 927.

[657] *T. P. O' Brien* and *J. W. O' Laughlin*," Gas chromatography of ternary complexes of manganese(II), iron(II), cobalt(II) and nickel(II) with hexafluoroacetylacetone and di-n-butyl sulfoxide", Talanta **23** (1976) 805.

[658] *H. Okabe, P. L. Splitstone,* and *J. J. Ball*, "Ambient and source sulphur dioxide detector based on a fluorescence method", J. Air Pollut. Contr. Ass. **23** (1973) 514.

[659] *J. W. O'Laughlin* and *T. P. O'Brien,* "Synergic extraction of zinc, cadmium, and lead with hexafluoroacetylacetone and di-n-butyl sulphoxide and tri-n-butyl phosphate and the gas chromatography of the zinc adduct", Talanta **22** (1975) 587.

[660] *G. Parissakis* and *B. Iatridis,* "GLC determination of silicon in alloys after chlorination with carbon tetrachloride", Chromatographia **10** (1977) 37.

[661] *G. Parissakis, D. Vranit-Piscon,* and *J. Kohtoyannakos,* "Gas chromatographic separation of metal chlorides", Z. Anal. Chem. **254** (1971) 188.

[662] *E. Patsalides, B. J. Stevenson,* and *S. Dilli,* "Gas chromatography of transition metal derivatives of the sulphur analogues of 4,4′-(ethane-1,2-salicyldiimino)-bis(pent-3-en-2-one)", J. Chromatogr. **173** (1979) 321.

[663] *I. B. Peetre,* "Gas chromatographic investigation of organometallic compounds and their carbon analogues II. Improved method for calculation of retention indices of tetraalkoxysilanes", J. Chromatogr. **88** (1974) 311.

[664] *I. B. Peetre,* "Gas chromatographic investigation of organometallic compounds and their carbon analogues II. A study of Kovats' retention indices for tetraalkoxysilanes containing bonded alkoxy groups", J. Chromatogr. **90** (1974) 35.

[665] *S. N. Pennington,* "Determination of oxidizing agents by gas chromatography-iodometry", J. Chromatogr. **36** (1968) 400.

[666] *R. E. Pescar* and *C. H. Hartmann,* "Automated gas chromatographic analysis of sulfur pollutants", J. Chromatogr. Sci. **11** (1973) 492.

[667] *C. S. G. Phillips* and *P. L. Timms,* "Some applications of gas chromatography in inorganic chemistry", Anal. Chem. **35** (1963) 505.

[668] *O. Pitak,* "Gas chromatography of inorganic fluorine compounds I. Gas chromatography of volatile, corrosive fluorides", Chromatographia **2** (1969) 304.

[669] *O. Pitak,* "Gas chromatography of inorganic fluorine compounds II. Investigation about efficiency of gas liquid chromatography for separation of volatile, corrosive inorganic fluorides", Chromatographia **2** (1969) 462.

[670] *O. Pitak,* "Gas chromatography of inorganic fluorine compounds III. Separation of corrosive inorganic fluorine compounds by means of gas-adsorption chromatography", Chromatographia **3** (1970) 29.

[671] *F. Pollard, G. Nickless,* and *P. C. Uden,* "Chromatographic studies of silicone compounds I. Use of a flame ionisation gauge in determination of retention data", J. Chromatogr. **14** (1964) 1.

[672] *G. Pommier* and *G. Guiochon*, "Gas chromatographic analysis of metal carbonyls", J. Chromatogr. Sci. **8** (1970) 486.

[673] *C. F. Poole, N. J. Evans,* and *D. G. Wibberley,* "Determination of selenium in bilogical samples by gas-liquid chromatography with electron-capture detection", J. Chromatogr. **136** (1977) 73.

*[674] *A. Przyjazny* and *R. Staszewski,* "Hochdruck-Gas-Chromatographie und Flüssigkeits-Chromatographie im superkritischen Zustand", Z. Anal. Chem. **272** (1974) 52.

[675] *A. Purer,* "A. procedure for analysis of impurities in grade-A helium in the parts per billion range", J. Gas Chromatogr. **3** (1965) 165.

[676] *R. C. Putnam* and *H. Pu,* "Retention indices of organotins II", J. Gas Chromatogr. **3** (1965) 289.

[677] *A. Radecki et al.,* "Gas-liquid chromatographic determination of zinc, copper, and nickel in marine bottom sediments", H. Chromatogr. **151** (1978) 259.

[678] *B. Radzink et al.,* "Determination of alkyl lead compounds in air by gas chromatography and atomic absorption spectrometry", Anal. Chim. Acta **105** (1979) 255.

[679] *G. Rambeau,* "Determination of ammonia in a mixture of hydrogen and nitrogen by gas-solid chromatography", Chim. Anal. **52** (1970) 774.

[680] *D. C. Reamer, W. H. Zoller,* and *T. C. O'Haver,* "Gas chromatograph-microwave plasma detector for the determination of tetraalkyllead species in the atmosphere", Anal. Chem. **50** (1978) 1449.

[681] *R. Rezl* and *J. Janak,* "Elemental analysis by gas chromatography", **81** (1973) 233.

[682] *W. A. Robbins* and *J. A. Caruso,* "Multielement analysis of selected elemental hydrides by chromatographically coupled plasma emission spectrometry", J. Chromatogr. Sci. **17** (1979) 360.

[683] *J. W. Robinson et al.,* "The development of a gas chromatograph-furnace atomic absorption combination for the determination of organic lead compounds. Atomization processes in furnace atomizers", Anal. Chim. Acta **92** (1977) 321.

[684] *J. A. Rodriguez-Vazquez,* "Quantitative inorganic analysis by gas chromatography", Anal. Chim. Acta **73** (1974) 1.

[685] *J. A. Rodriguez-Vazquez,* "Gas-chromatographic determination of organomercury(II) compounds", Talanta **25** (1978) 299.

[686] *P. Ronkainen, J. Denslow,* and *O. Leppänen,* "The gas chromatographic analysis of some volatile sulfur compounds", J. Chromatogr. Sci. **11** (1973) 384.

[687] *R. T. Ross* and *M. T. Shafik,* "The determination of chromium in human urine by gas chromatography using a flame photometric detector with a 425.4 nm filter", J. Chromatogr. Sci. **11** (1973) 46.

[688] *W. D. Ross,* "Detection of metal chelates in gas liquid chromatography by electron capture", Anal. Chem. **35** (1963) 1596.

[689] *W. D. Ross* and *R. E. Sievers,* "Rapid ultra-trace determination of beryllium by gas chromatography", Talanta **15** (1968) 87.

[690] *W. D. Ross* and *R. E. Sievers,* "Quantitative analysis for trace chromium in ferrous alloys by electron capture gas chromatography", Anal. Chem. **41** (1969) 1109.

[691] *W. D. Ross* and *G. Wheeler*, "Quantitative determination of chromium(III) hexafluoroacetylacetonate by gas chromatography", Anal. Chem. **36** (1964) 266.

[692] *W. D. Ross, R. E. Sievers,* and *G. Wheeler,* "Quantitative ultratrace analysis of mixtures of metal chelates by gas chromatography", Anal. Chem. **37** (1965) 598.

[693] *W. D. Ross et al.,* "Analysis for aqueous nitrates and nitrites and gaseous oxides of nitrogen by electron capture gas chromatography", J. Chromatogr. **112** (1975) 719.

[694] *B. A. Rudenko, M. A. Baidarovtseva,* and *M. A. Agaeva,* "Elution chromatography with vapours as mobile phases", Zh. analit. Khim. **30** (1975) 1191 – Anal. Abstr. **30** (1976) 6J16.

*[695] *N. K. Rudnevskii et al.,* "Atomic absorption spectroscopy as detector in gas chromatography (in the determination of antimony and gallium)", Anal. Abstr. **30** (1976) 3B65.

[696] *J. Rudolph* and *K. Bächmann,* "Inorganic gas chromatography I. Principles and optimization of isothermal separation conditions", Chromatographia **10** (1977) 731.

[697] *J. Rudolph* and *K. Bächmann,* "A gas chromatograph apparatus for the investigation and separation of radioactively labelled inorganic compounds at high temperatures", Mikrochim. Acta I (1979) 477.

[698] *J. Rudolph et al.,* "Inorganic gas chromatography at high temperatures", Mikrochim. Acta I (1978) 471.

[699] *H. A. Rüssel,* "Determination of halide ions by gas chromatography", Angew. Chem. **82** (1970) 392; Angew. Chem. Int. Edit. Engl. **9** (1970) 374.

[700] *H. A. Rüssel,* "Gas chromatographic determination of fluoride in organic materials and body fluids", Z. Anal. Chem. **252** (1970) 143.

[701] *H. Rüssel* and *G. Tölg,* "Anwendung der Gaschromatographie zur Trennung und Bestimmung anorganischer Stoffe", Fortschr. chem. Forsch. **33** (1972) 1.

[702] *J. J. Ryan* and *J. A. Dupont,* "Identification and determination of the major acids of fruit juices and wines", J. Agr. Food Chem. **21** (1973) 45.

[703] *T. R. Ryan* and *C. R. Hastings Vogt,* "Determination of physiological levels of Cr(III) in urine by gas chromatography", J. Chromatogr. **130** (1977) 346.

[704] *T. Sakamoto, H. Kawaguchi,* and *A. Mizuike,* "Determination of traces of copper and aluminium in zinc by gas chromatography with the microwave plasma detector", J. Chromatogr. **121** (1976) 383.

[705] *T. Sakamoto, H. Kawaguchi,* and *A. Mizuike,* "Gas chromatography of arsenic in soil using the emission spectrometric detector", Japan Analyst **25** (1976) 81.

[706] *T. Sakamoto et al.,* "Gas chromatography of traces of beryllium using the microwave plasma detector", Japan Analyst **25** (1976) 85.

[707] *T. Sakano, Y. Hori,* and *Y. Tomari,* "Gas chromatographic determination of trace amounts of water in organic solvent containing active chlorine and hydrogen chloride", J. Chromatogr. Sci. **14** (1976) 501.

[708] *M. Saleem et al.,* "Complex formation of antimony trichloride in sodium tetrachloroferrate(III) with excess sodium chloride and determination of thermodynamic parameters by gas chromatography", Chromatographia **8** (1975) 699.

[709] *J. Savory, M. T. Glenn,* and *J. A. Ahlstrom,* "Determination of physiological levels of chromium in serum by gas chromatography", J. Chromatogr. Sci. **10** (1972) 247.

[710] *J. Savory, P. Mushak,* and *F. W. Sunderman,* "Gas chromatographic determination of chromium in serum", J. Chromatogr. Sci. **7** (1969) 674.

[711] *J. Savory et al.,* "Gas chromatographic micro determination of chromium in biological materials", Anal. Chem. **42** (1970) 294.

*[712] *M. L. Sazonov et al.,* "Gas-chromatographic determination of germanium in coal", Anal. Abstr. **17** (1969) 881.

[713] *F. P. Scaringelli* and *K. H. Rehmke,* "Determination of atmospheric concentrations of sulfuric acid aerosol by spectrometry, coulometry, and flame photometry", Anal. Chem. **41** (1969) 707.

[714] *D. A. Segar,* "Flameless atomic absorption-gas chromatography", Anal. Lett. **7** (1974) 89.

[715] *J. A. Semlyen* and *C. S. G. Phillips,* "The compilation and correlation of retention data for alkylsilanes, germanes, digermanes, and borazoles and some of their hydrocarbon analogues", J. Chromatogr. **18** (1965) 1.

[716] *F. A. Serravallo* and *T. H. Risby,* "Metal selective microwave emission detector for gas chromatography", J. Chromatogr. Sci. **12** (1974) 585.

[717] *F. A. Serravallo* and *T. H. Risby,* "Effect of doping gases on microwave-induced emissive spectrometric detectors for gas chromatography", Anal. Chem. **47** (1975) 2141.

[718] *F. A. Serravallo* and *T. H. Risby,* "A metal selective microwave emission detector for gas chromatography", J. Chromatogr. Sci. **12** (1975) 585.

[719] *J. Sevcik,* "Detectors in gas chromatography", Journal of chromatography library Vol. 4, Elsevier Amsterdam (1976).

[720] *V. D. Shatz et al.,* "Chromatography of organometallic and organometalloidal derivatives of amino alcohols I. Retention indices of silatranes", J. Chromatogr. **174** (1979) 83.

[721] *M. Shaviat,* "Determination of dialkylmercury compounds by gas chromatography", J. Chromatogr. Sci. **17** (1979) 527.

[722] *T. Shigematsu, M. Matsui,* and *K. Utsunomiya,* "Gas chromatography of diisobutyrylmethane metal chelates", Bull. Inst. Chem. Res. Kyto Univ. **46** (1968) 256.

[723] *T. Shigematsu, M. Matsui,* and *K. Utsunomiya,* "Gas chromatography of rare-earth-metal chelates of pivaloyltrifluoroacetone", Bull. Chem. Soc. Japan **41** (1968 763, **42** (1969) 1278.

[724] *Y. Shimoishi,* "The determination of selenium in sea water by gas chromatography with electron capture detection", Anal. Chem. Acta **64** (1973) 465.

[725] *Y. Shimoishi,* "The gas-chromatographic determination of selenium(IV) and total selenium in milk, milk products, and albumin with 1,2-diamino-4-nitrobenzene", Analyst **101** (1976) 298.

[726] *Y. Shimoishi,* "Some 1.2-diaminobenzene derivatives as reagents for gas chromatographic determination of selenium with an electron-capture detector", J. Chromatogr. **136** (1977) 85.

[727] *Y. Shimoishi* and *K. Toei,* "Gas chromatographic determination of ultramicro amounts of selenium in pure sulphuric acid", Talanta **17** (1970) 165.

[728] *Y. Shimoishi* and *K. Toei,* "The gas chromatographic determination of seleniumIV and total selenium in natural waters with 1,2-diamino-3,5-dibromobenzene", Anal. Chim. Acta **100** (1978) 65.

[729] *S. T. Sie, J. P. A. Bleumer,* and *G. W. A. Rijnders,* "Gas-chromatographic separation of inorganic chlorides and its application to metal analysis II. Determination of silicon in iron and steel", Separ. Sci. **2** (1967) 645.

[730] *S. T. Sie, J. P. A. Bleumer,* and *G. W. A. Rijnders,* "Gas-chromatographic separation of inorganic chlorides and its application to metal analysis, III. Determination of silicon and tin in non-ferrous alloys", Separ. Sci. **3** (1968) 165.

[731] *R. F. Sieck* and *C. V. Banks,* "Gas chromatography of volatile mixed-ligand complexes of the lanthanides", Anal. Chem. **44** (1972) 2307.

[732] *R. Sieck et al.,* "Determination of uranyl and thoriumIV by gas chromatography of volatile mixed-ligand complexes", Anal. Chem. **43** (1971) 913.

[733] *R. E. Sievers, J. W. Connolly,* and *W. D. Ross,* "Metal analysis by gas chromatography of chelates of heptafluorodimethyloctanedione", J. Gas Chromatogr. **5** (1967) 241.

[734] *R. E. Sievers et al.,* "Microanalysis by gaschromatography: Determination of metals in lunar dust and rock", VI. Intern. Symp. Mikrochem. Graz Vol. B (1970) 247.

[735] *L. J. Soderquist* and *D. G. Crosby,* "Determination of triphenyltin hydroxide and its degradation products in water", Anal. Chem. **50** (1978) 1435.

[736] *D. N. Sokolov et al.*, "Sorption effects in columns in the gas chromatography of copper trifluoroacetylacetonate", J. Chromatogr. **155** (1978) 241.
[737] *E. J. Sowinski* and *I. H. Suffet,* "Characterization of a Melpar flame photometric-gas chromatographic system for application to boron hydrides", J. Chromatogr. Sci. **9** (1971) 632.
[738] *E. J. Sowinski* and *I. H. Suffet,* "Programmed temperature gas chromatography of boron hydrides", Anal. Chem. **44** (1972) 2237.
[739] *E. J. Sowinski* and *I. H. Suffet,* "Gas chromatographic detection and confirmation of volatile boron hydrides at trace levels", Anal. Chem. **46** (1974) 1218.
[740] *S. R. Srinivasan et al.,* "Determination of sulphate in glycosamine glycans by gas chromatography", Anal. Biochem. **35** (1970) 398.
[741] *B. A. Schaefer,* "Analysis of inorganic sulfur compounds by flame ionisation detector", Anal. Chem. **42** (1970) 448.
[742] *B. A. Schaefer,* "Response of the flame ionization detector to oxygen and nitrous oxide", J. Chromatogr. Sci. **10** (1972) 110.
[743] *B. A. Schaefer,* "Response of the flame ionization detector to water and formic acid", J. Chromatogr. Sci. **13** (1975) 86.
[744] *B. A. Schaefer* and *D. M. Douglas,* "Response of the flame ionization detector to carbon monoxide and carbon dioxide", J. Chromatogr. Sci. **9** (1971) 612.
[745] *M. L. Schafer et al.,* "Method for the estimation of methylmercuric compounds in fish", J. Agric. Food Chem. **23** (1975) 1079.
[746] *H. Schnecko, C. D. Eisenbach,* and *W. Kearn,* "Indirect qualitative determination of organolithium compounds by gas chromatography", J. Chromatogr. Sci. **14** (1976) 219.
[747] *H. Schneider* and *R. Neeb,* "Gas-chromatographische Elementanalyse über Di(trifluoräthyl)dithiocarbamatochelate IV. Die Verteilung der Di(trifluoräthyl)dithiocarbamate des CuII, NiII, CoIII, FeIII, BiIII, PbII und SbIII zwischen wäßrigen Lösungen und Chloroform. Extraktive gas-chromatographische Simultanbestimmung von Blei und Wismut", Z. Anal. Chem. **293** (1978) 11.
[748] *I. Schuphan, K. Ballschmiter,* and *G. Tölg,* "Polychlorierte Verbindungen und ihre Verwendung zur gaschromatographischen Bestimmung von Elementen im ng-Bereich I. Polychlorierte Xanthogenate und ihre Anwendung zur Nickel-Bestimmung", Z. Anal. Chem. **255** (1971) 116.
[749] *J. E. Schwarberg* and *R. W. Moshier,* "Feasibility of gas-liquid chromatography for quantitative determination of Al(III), In(III), Ga(III) and Be(II)-trifluoroacetylacetonates", Talanta **11** (1964) 1213.

[750] *J. E. Schwarberg, R. E. Sievers,* and *R. W. Moshier,* "Gas chromatographic and related properties of the alkaline earth chelates with 2,2,6,6-tetramethyl-3,5-heptanedione", Anal. Chem. **42** (1970) 1828.

[751] *G. Schwedt,* "Chromatography in inorganic trace analysis", Topics in Current Chemistry **85** (1979) 159.

[752] *G. Schwedt* and *H. A. Rüssel,* "Gas-chromatographische Bestimmung von Arsen als Triphenylarsan", Chromatographia **5** (1972) 242.

[752a] *G. Schwedt* and *H. A. Rüssel,* "Gas-Chromatographie von Tl, Se, Te, Hg, As, Sb, Bi, Sn als Phenylverbindungen", Z. Anal. Chem. **264** (1973) 301.

[753] *A. Steffen* and K. Bächmann, "Gas chromatographic study of volatile oxides and hydroxides of Tc, Re, Ru, Os and Ir I. Investigation of isothermal gas chromatography", Talanta **25** (1978) 551.

[754] *W. I. Stephen,* "Gas chromatography of metal chelates", Proc. Soc. Anal. Chem. **9** (1972) 137.

[755] *W. J. Stephen, I. J. Thomson,* and *P. C. Uden,* "Gas-chromatography of transition-metal monothioacetylacetonates", Chem. Commun. (1969) 269.

[756] *J. R. Stetter, D. R. Rutt,* and *K. F. Blurton,* "Portable gas chromatograph and electrochemical detector for carbon monoxide", Anal. Chem. **48** (1976) 924.

[757] *R. K. Stevens, A. E. O'Keeffe,* and *G. C. Ortman,* "Absolute calibration of a flame photometric detector to volatile sulfur compounds at sub-part-per-million-levels", Environ. Sci. Technol. **3** (1969) 652.

[758] *T. Stijve* and *E. Cardinale,* "Rapid determination of selenium in various substrates by electron capture gas-liquid chromatography", J. Chromatogr. **109** (1975) 239.

[759] *J. Tadmor,* "Gas chromatography in inorganic chemistry", Chromatogr. Rev. **5** (1963) 223.

[760] *F. Taitiro, K. Tooru,* and *M. Shigeo,* "Gas chromatography of metal chelates with carrier gas containing ligand vapour", Talanta **18** (1971) 429.

[761] *Y. Talmi,* The rapid sub-picogram determination of volatile organomercury compounds by gas chromatography with a microwave emission spectrometric detector system", Anal. Chim. Acta **74** (1975) 107.

[762] *Y. Talmi* and *A. W. Andren,* "Determination of selenium in environmental samples using GC with a microwave emission spectrometric detector system", Anal. Chem. **46** (1974) 2122.

[763] *Y. Talmi* and *D. T. Bostick,* "Determination of alkylarsonic acids in pesticide and environmental samples by gas chromatography with a microwave emission spectrometric detection system", Anal. Chem. **47** (1975) 2145.

[764] *Y. Talmi* and *V. E. Norvell,* "Determination of arsenic and antimony in environmental samples using gas

chromatography with a microwave emission spectrometric system", Anal. Chem. **47** (1975) 1510.

[765] *M. Tanaka, T. Shono,* and *K. Shiura,* "Thermogravimetric analysis and gas chromatography of rare earth chelates of trifluoroacetylpivalolylmethane", Anal. Chim. Acta **43** (1968) 157.

[766] *K. Tanikawa,* "Organometallic compounds X. Behaviour of β-diketone-metal chelates in gas chromatography and effect of chelate stability on detection sensitivity", Japan Analyst **25** (1976) 721.

[767] *K. Tanikawa, H. Ochi,* and *K. Arakawa,* "Gas chromatographic analysis of metal pivaloyltrifluoroacetonates", Japan Analyst **19** (1970) 1669.

[768] *R. L. Tanner, R. Fajer,* and *J. Gaffney,* "Determination of parts-per-billion concentrations of aqueous nitrate by derivatization gas chromatography with electron capture detection", Anal. Chem. **51** (1979) 865.

[769] *A. Tavlaridis* and *R. Neeb,* "Gas-chromatographische Multielementanalyse mit einfachen und fluorierten Diäthyldithiocarbamaten", Z. Anal. Chem. **282** (1976) 17.

[770] *A. Tavlaridis* and *R. Neeb,* "Gas-chromatographische Elementanalyse über Di(trifluoräthyl)dithiocarbamatochelate II. Das gas-chromatographische Verhalten einiger Di(trifluoräthyl)dithiocarbamatochelate", Z. Anal. Chem. **292** (1978) 199.

[771] *A. Tavlaridis* and *R. Neeb,* "Dialkyldithiocarbamate als Reagentien zur gas-chromatographischen Bestimmung von Metallen I. Thermisches Verhalten, Dampfdrucke und Gas-Chromatographie einiger Dialkyldithiocarbamatochelate", Z. Anal. Chem. **293** (1978) 211.

[772] *M. L. Taylor* and *E. J. Arnold,* "Ultratrace determination of metals in biological specimens – quantitative determination of beryllium by gas-chromatography", Anal. Chem. **43** (1971) 1328.

[773] *M. L. Taylor, E. L. Arnold,* and *R. E. Sievers,* "Rapid microanalysis for beryllium in biological fluids by gas chromatography", Anal. Lett. **1** (1968) 735.

[774] *K. Tesarrik* and *M. Krejci,* "Chromatographic determination of carbon monoxide below the 1 ppm level", J. Chromatogr. **91** (1974) 539.

[775] *J. W. Tesch, W. R. Rehg,* and *R. E. Sievers,* "Microdetermination of nitrates and nitrites in saliva, blood, water, and suspended particulates in air by gas chromatography", J. Chromatogr. **126** (1976) 743.

[776] *W. L. Thornsberry,* "Isothermal gas chromatographic separation of carbon dioxide, carbon oxysulfide, hydrogen sulfide, carbon disulfide, and sulfur dioxide", Anal. Chem. **43** (1971) 452.

[777] *I. Tohyama* and *K. Otazai,* "High-temperature gas-chromatography of some inorganic chlorides", Z. Anal. Chem. **262** (1972) 346.

[778] *I. Tonyama* and *K. Otozai*, "Gas chromatography of $SiCl_4$, $GeCl_4$, $TiCl_4$", Z. Anal Chem. **271** (1974) 117.

[779] *B. L. Tonge*, "The gas chromatographic analysis of butyl-, octyl- and phenyltin halides," J. Chromatogr. **19**, (1965) 182.

[780] *D. A. Tourres*, "Trace determination of hydrogen sulfide and sulfur dioxide by rapid, on-line gas-chromatography", Chromatographia **5** (1972) 441.

[781] *S. Tsalas* and *K. Bächmann*, "Inorganic gas chromatography – the separation of volatile chlorides by thermochromatography combined with complex formation", Anal. Chim. Acta **98** (1978) 17.

[782] *G. V. Isitsishvili*, et al. "Chromatographic properties of type-X zeolites containing alkali-earth metal cations", Chromatographia **6** (1973) 366.

[783] *P. C. Uden* and *K. Blessel*, "Resolution of volatile N, N'-2,3-butylenebis(trifluoroacetylacetone iminato)copper(II), -nickel(II), and -palladium(II) chelates by gas-liquid chromatography", Inorg. Chem. **12** (1973) 352.

[784] *P. C. Uden D. E. Henderson*, "Non-fluorinated tetradentate β-ketoamines as derivatizing ligands for the gas chromatographic analysis of Cu(II), Ni(II) and Pd(II)", J. Chromatogr. **99** (1974) 309.

[785] *P. C. Uden* and *D. E. Henderson*, "Determination of metals by chromatography of metal complexes. A. review", Analyst **102** (1977) 889.

[786] *P. C. Uden* and *C. R. Jenkins,* "Adsorption and displacement effects in the gas-chromatography of metal β-diketonates", Talanta **16** (1969) 893.

[787] *P. C. Uden* and *B. A. Waldman*, "Gas chromatography of transition metal salicylaldimine complexes", Anal. Lett. **8** (1975) 91.

[788] *P. C. Uden, D. E. Henderson,* and *C. A. Burgett*, "Trace copper and nickel chelate detection by electron capture gas chromatography", Anal Lett. **7** (1974) 807.

[789] *P. C. Uden, D. E. Henderson,* and *A. Kamalizad*, "Electron capture gas chromatographic determination of copper and nickel β-ketoamine chelates", J. Chromatogr. Sci. **12** (1974) 591.

[790] *E. Uhlemann et al.* "Synthese, Koordinationschemie und analytische Anwendung von Monothio-β-diketonen", Z. Chemie **11** (1971) 401.

[791] *K. Utsonmiya*, "Gas chromatography of rare earth chelates of isobutyrylpivaloylmethane", Anal. Chim. Acta **59** (1972) 147.

[792] *K. Utsonomiya* and *T. Shigematsu*, "Thermogravimetric and gas-chromatographic study of neodymium, gadolinium and erbium β-diketone chelates", Anal. Chim. Acta **58** (1972) 411.

[793] *J. C. Valentour, V. Aggarwal* and *I. Sunshine*," Sensitive gas chromatographic determination of cyanide", Anal. Chem. **46** (1974) 924.

[794] *H. Venning, W. E. Bachmann,* and *D. M. Wilkinson*, "Preparation and gas chromatography of cobaltIII and rutheniumIII fluoroacetylacetonates", J. Gas Chromatogr. **5** (1967) 248.

[795] *B. Vierkorn-Rudolph et al.* "Determination of small concentration of HCl by derivative formation – a problem and how it might be solved", Talanta **26** (1979) 755.

[796] *J. Warren* and *A. G. Smeeth*, "Thermal studies and gas chromatography of some metal β-diketonates", Proc. Soc. Anal. Chem. **9** (1972) 194.

[797] *J. O. Watts et al.* "Simplified method for gas-liquid chromatographic determination of methylmercury in fish and shellfish", J. Assoc. Off. Anal. Chem. **59** (1976) 1226.

[798] *J. Wells* and *G. Cimbura*, "The determination of elevated bromide levels in blood by gas chromatography", J. Forensic. Sci. **18** (1973) 437.

[799] *P. M. Wiese* and *R. H. Hanson*, "Gas chromatographic analysis of aqueous phosphate by reaction gas chromatography", Anal. Chem. **44** (1972) 2393.

[800] *D. L. Windsor* and *M. Bonner Denton*, "Elemental analysis of gas chromatographic effluents with an inductively coupled plasma", J. Chromatogr. Sci. **17** (1979) 492.

[801] *M. K. Withers*, "Reduction of tailing due to aluminum tubing used for gas chromatographic columns", J. Chromatrogr. **80** (1973) 124.

[801a] *J. Wilke, A. Losse,* and *L. Sackmann*, "Contribution to the gas chromatography of tetrachlorides", $SiCl_4$, $SnCl_4$ and $TiCl_4$", J. Chromatogr. **18** (1965) 482.

[802] *W. R. Wolf*, "Coupled gas chromatography-atomic absorption spectrometry for the nanogram determination of chromium", Anal. Chem. **48** (1976) 1717.

[803] *W. R. Wolf*, "Coupled gas chromatography-atomic absorption spectrometry", J. Chromatogr. **134** (1977) 159.

[804] *W. R. Wolf, R. E. Sievers,* and *G. H. Brown*, "Vapour pressure measurements and gas chromatographic studies of the solution thermodynamics of metal β-diketonates", Inorg. Chem. **11** (1972) 1995.

[805] *W. R. Wolf* et al. "Determination of chromium and beryllium at the picogram level by gas chromatography-mass spectrometry", Anal. Chem. **44** (1972) 616.

[806] *H. F. F. Wu et al.* "Determination of silicate anions by gas chromatographic separation and mass spectrometric identification of their trimethylsilyl derivatives", J. Chromatogr. **48** (1970) 515.

[807] *I. Youdina, Y. Youzhelevski* and *K. Sakodyasky*, "Chromatographic analysis of methyl(propyl)dimethylcyclosiloxanes", J. Chromatogr. **38** (1968) 240.

[808] *J. W. Young* and *G. D. Christian*, "Gas-chromatographic determination of selenium", Anal. Chim. Acta **65** (1973) 127.

[809] *F. M. Zado* and *R. S. Juvet,* "A new selective-nonselective flame photometric detector for gas chromatography", Anal. Chem. **38** (1966) 569.

[810] *P. Zarnegar* and *P. Mushak,* "Quantitative measurements in inorganic mercury and organomercurials in water and biological media by gas-liquid chromatography", Anal. Chim. Acta **69** (1974) 389.

[811] *E. N. Zerba, M. A. Ruveda,* and *R. Libertella,* "Organophosphorus liquid phases: characterization and application to the gas chromatographic analysis of organophosphorus compounds", J. Chromatogr. **108** (1975) 380.

[812] *E. Zielinski,* "Use of argon ionisation detector with 63-Ni source for chromatographic determination of hydrogen sulfide, sulfur dioxide, and carbon oxysulfide", Chem. Anal. **14** (1969) 521 – Z. Anal. Chem. **254** (1971) 378.

[813] *I. Zvara et al.,* "Thermochromatography of bromides. A proposed technique for the study of transactinide element chemistry", J. Chromatogr. **103** (1975) 77.

[814] *T. S. Zvarova* and *I. Zvara,* "Separation of rare-earth elements by gas chromatography of their chlorides", J. Chromatogr. **441** (1969) 604.

V Indexes

1 Chromatography of Elements – Documentation

This documentation covers work on quantitative analysis following chromatographic separation and on the application of chromatography to inorganic analysis of elements and their compounds.

Publications concerned with the general chromatographic behavior of several elements or compounds in particular systems are accessible with the aid of the method (key word) index.

Abbreviations

TLC, thin-layer chromatography (includes paper chromatography); CC, column chromatography (includes HPLC); GC, gas chromatography.

The number of the publication in Chapter IV is given in each case.

2 Method (Key Word) Index

This index covers Chapters I and II, as well as Sections 3 and 4 of Chapter III. Reference to "Applications in Inorganic Trace Analysis" is facilitated by the tabular presentation of material.

Page numbers in italics refer to tables and those in bold print to figures.

Walter Jennings

Comparisons of Fused Silica and other Glass Columns in Gas Chromatography

1981, approx. 140 pp., numerous figures and tables, hard cover,
DM 43,–/$20.00 (tent.)
ISBN 3-7785-0729-X

This highly readable and timely book deals with the various types of glass columns used in gas chromatography. It opens with a detailed account of the construction materials soda-lime, borosilicate, lead, fused quarts and fused silica glasses, which is followed by a description of the pretreatment of the capillary. The succeeding chapter is concerned with methods of evaluating inertness and describes methods of testing uncoated capillaries and coated columns. The physical characteristics of glass columns are also considered and the advantages conferred by column flexibility are presented at length. The monograph concludes with cost comparisons of glass and fused silica columns and numerous examples of practical applications.

Contents

Introduction · Materials of construction · Pretreatment of the capillary · Evaluation of column inertness · Physical characteristics of the column · Cost comparisons of glass and fused silica columns · Application examples.